In 1985, Buckminsterfullerene (fullerene-60) C_{60} was discovered serendipitously during graphite laser vaporization experiments designed to simulate the chemistry in a red giant carbon star. The molecule was isolated for the first time in macroscopic amounts in 1990, a breakthrough which triggered an explosion of research into its chemical and physical properties. C_{60} has already exhibited a wide range of novel phenomena which hold promise for exciting applications. Whether or not these applications actually materialize, Buckminsterfullerene possesses a beauty and elegance that has excited the imaginations of laymen and scientists alike. It seems almost impossible to comprehend how the existence of the third well-characterized allotrope of carbon could have evaded discovery until virtually the end of the twentieth century. In October 1992 a Discussion Meeting of the Royal Society entitled 'A Post-Buckminsterfullerene View of the Chemistry, Physics and Astrophysics of Carbon' organized by H. W. Kroto, A. L. MacKay, G. Turner and D. R. M. Walton, was held to celebrate this exciting advance. The papers, presented by those who played key roles in the discovery and by others who are currently uncovering fascinating problems and the implications of this elegant molecule, are presented in this book.

The Fullerenes

The Fullerenes

New Horizons for the Chemistry, Physics and Astrophysics of Carbon

Edited by
H. W. Kroto and D. R. M. Walton

Published by the Press Syndicate of the University of Cambridge
The Pitt Building, Trumpington Street, Cambridge CB2 1RP
40 West 20th Street, New York, NY 10011–4211, USA
10 Stamford Road, Oakleigh, Melbourne 3166, Australia

First published by the Royal Society 1993 as Volume 343 Number 1667
of *Philosophical Transactions of the Royal Society*, A,

First published as a book by Cambridge University Press 1993

Printed in Great Britain at the University Press, Cambridge

A catalogue record for this book is available from the British Library

Library of Congress cataloguing in publication data available

ISBN 0 521 45917 6 paperback

UP

Contents

The evolution of the football structure for the C_{60} molecule: a retrospective

By E. Ōsawa

Department of Knowledge-based Information Engineering, Toyohashi University of Technology, Toyohashi 441, Japan

By chance in 1970, we conjectured the possibility of the football-shaped C_{60} molecule, now known as buckminsterfullerene, while considering superaromatic molecules having three-dimensional π-electron delocalization. A translation of the original description, initially written in Japanese, is given. The processes leading to scientific discoveries are analysed in the light of our missed opportunity.

1. Introduction

The timescale of scientific and technological advance is becoming shorter and shorter in modern society, partly as a consequence of the rapid advance of technology and improving information transfer. It is no wonder then that the time has come to look back and discuss the future of fullerene science after less than a decade since its discovery by Kroto *et al.* (1985) and after only two years since it was isolated by Krätchmer *et al.* (1990). The purpose of this paper is to recount the story of original early proposal of the football-shaped C_{60} molecule back in 1970, and refer to other interesting 'prehistoric' events and analyse the process of scientific discovery.

2. Background

In the 1960s and 1970s, non-benzenoid aromatics were favourite targets for organic chemists. There was a prevailing dogma that aromaticity, due to the delocalization of π-electrons, is best realized in planar molecules. Everyone wished to constrain their molecules to be as planar as possible and for this reason aromaticity tacitly remained a two-dimensional concept. [18]Annulene (**1**), synthesized by Sondheimer *et al.* (1962), can be regarded as the masterpiece of planar aromaticity for its symmetric beauty (D_{6h}) and high level of π-electron delocalization. In view of the wide availability and its perfect aromaticity, however, benzene remains the archetypal superstar of aromatic molecules.

Our goal in those days was to find a new π-system more aromatic than king benzene. Then came the news of the corannulene (**6**) synthesis by Barth & Lawton (1966) and its bowl-shaped structure quickly aroused strong interest among us. As is mentioned below, analysis of its structure logically led us to conceive the extrapolation of its structure to a sphere so that we could envisage three-dimensional delocalization of the π-electrons. Fortuitously about that time my small son started

Printed in Great Britain

to play with a football and the pattern of the corannulene molecule was immediately and clearly recognizable in it. Careful study of the design of the ball soon led to the recognition that it was a truncated icosahedron.

The I_h-symmetric C_{60} molecule was first described by Ōsawa (1970), and the concept expanded the following year in the final chapter of Yoshida & Ōsawa (1971). The book was read widely by Japanese chemists. There was, however, little response to my proposal of the potential stability of the football-shaped molecule. In the meantime, for reasons mentioned below, we stopped working on aromaticity and engaged in new fields of research until the classic paper of Kroto *et al.* (1985) appeared. In 1986, O'Brien, then a graduate student with Smalley at Rice University, U.S.A., asked Yoshida for an English copy of this book and we produced a translation of pertinent portions (pp. 174–178) for him. This translation (slightly polished) is reproduced below.

3. Possibilities of superaromatic hydrocarbons

We temporarily define the term 'superaromaticity' as the lowering of energy that might occur when electrons delocalize over molecular orbitals on the surface of some three-dimensional surface of high symmetry. Is there any possibility of ever realizing such a phenomenon with hydrocarbons?

Let us take a look at the back cover of Cram and Hammond's *Textbook of organic chemistry*, where a number of 'dream molecules' are depicted. Truncated tetrahedrane, or heptacyclo[$5.5.0.0^{2,12}.0^{3,5}.0^{4,10}.0^{6,8}.0^{9,11}$]dodecane, $C_{12}H_{12}$ (**2**) appears to offer a possibility of its cyclopropane bonds with high p-character to interact over the surface of the molecule. Structure **2** has eight faces and is formally obtained by truncating four vertices of tetrahedron (**3**): it is also the target of synthesis by several research groups (Woodward 1970).

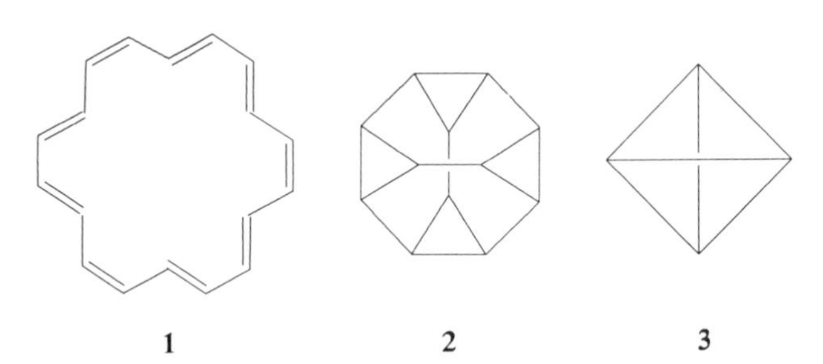

Another possibility of realizing three-dimensional (3D) aromaticity would be to consider electron delocalization, not among σ orbitals on the sphere, but through p_z orbitals directed perpendicular to the surface of sphere. If such conjugation is to be achieved over the surface of spheroidal skeleton composed of carbon atoms, then the sphere will have to be large, so that the p_z–p_z overlap is not reduced too much compared with that for a planar skeleton. If we follow the strategy of truncating a Platonic solid to produce a near-spheroidal structure, as mentioned above, the icosahedron (**4**) seems to provide a large enough skeleton for our purpose.

Truncation of a vertex of this solid produces a regular pentagon at that place and truncation of all twelve vertices gives a beautiful 32-faced solid, which is called the truncated icosahedron (**5**). The solid has the same design as that appearing on the surface of an official modern football. Study the ball yourself. Among the pentagons, hexagons are buried in a highly symmetric manner. The component polygons do not seem to be deformed very much from their planar form, and all the 90 edges have almost the same length. Then, the C_{60} molecule which results from replacing 60 vertices by sp^2-hybridized carbon atoms does not seem totally unrealistic.

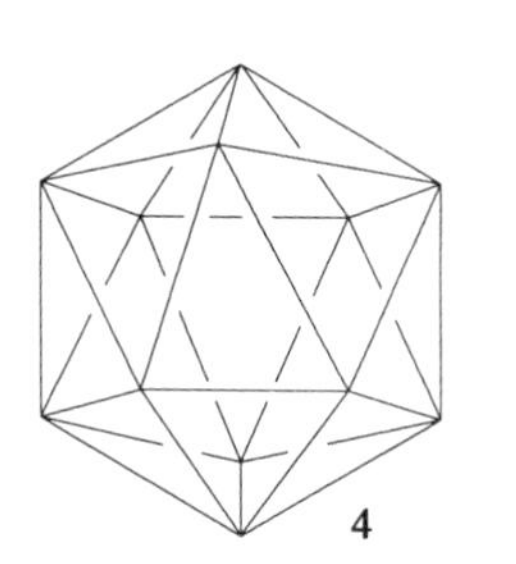

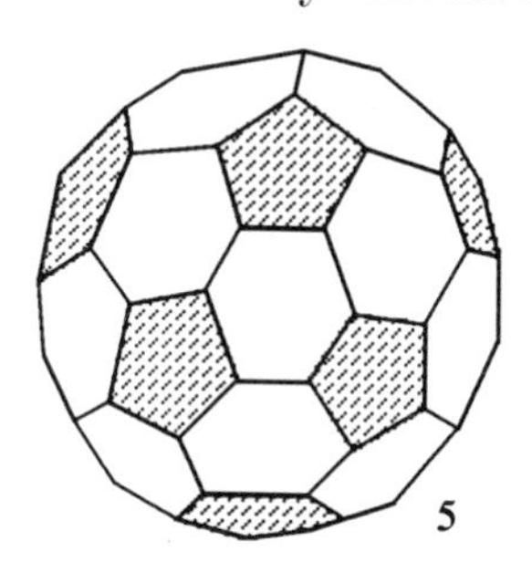

Interestingly enough, Barth & Lawton (1966, 1971) recently synthesized a molecule that corresponds to a part of the surface design of the football: dibenzo[ghi,mno]fluoranthene (**6**) with a five-membered ring surrounded by five six-membered ring. This hydrocarbon gives almost colourless, stable prismatic crystals melting at 268–269 °C and is named corannulene. According to X-ray analysis, the molecule is bowl-shaped (**7**) and indeed appears to correspond to a segment of structure **5**.

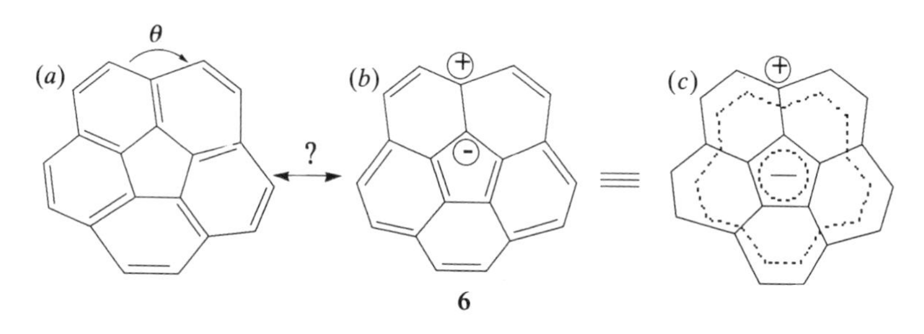

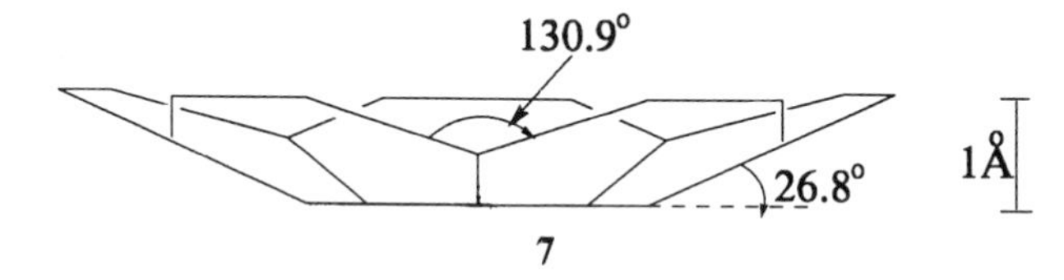

It should be noted, however, that planarity had not been excluded for corannulene until the X-ray analysis was performed. The molecule is a non-alternant hydrocarbon because it contains an odd-membered ring. Hence, the π-electron density distribution cannot be uniform in the ground state. One of such polar structures is the double Hückel aromatic structure **6***b*, *c* consisting of peripheral 14π and central 6π systems. The contribution of resonance as shown in **6***c* will be at its maximum when the whole molecule is planar like coronene (**8**). Notwithstanding, the planar corannulene will have huge angle strain. If we assume that all C–C bonds are 1.40 Å in length and the central ring is regular pentagon for the planar corannulene, then the angle θ shown in **6**a will be as large as 144°. Out-of-plane deformation of five bonds extending from the pentagon may release some of the angle strain. The deformation would be, however, as large as 38° if the standard value of 120° is to be restored for θ (**9**). Then the neighbouring benzene rings would no longer be coplanar and the stabilization energy, by delocalizing the π-electrons, would be greatly reduced.

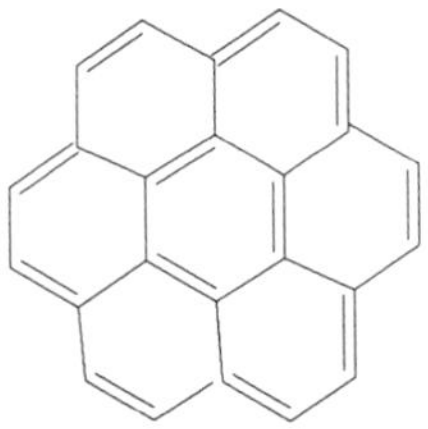

120° 38°

8 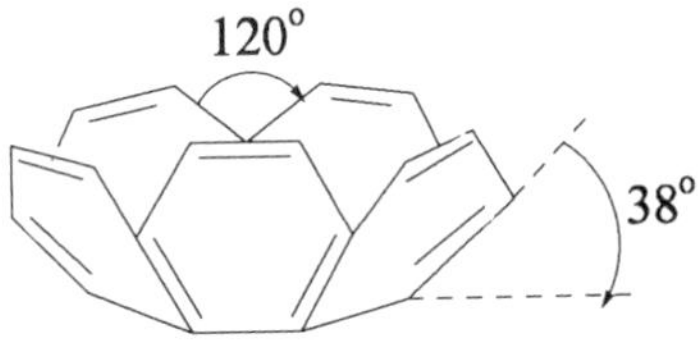 **9**

At this point, we notice that the corannulene molecule provides an ideal model for addressing a very interesting problem: how much aromaticity is lost when a molecule is distorted from the strongly aromatic planar structure to a nonplanar structure? The answer to this question must be a key to the possible existence of *superaromaticity* in the C_{60} molecule (**5**).

Let us look again at the X-ray structure of corannulene (**7**). The observed angle between the pentagon and the hexagons is 26.8° in average. Therefore, considerable strain must remain in the skeleton. The observed angle θ between the six-membered rings is 130.9°. The actual structure of the corannulene molecule is a shallow bowl (**7**) somewhere between the planar structure (**6**) and the deep-bowl form (**9**). The NMR chemical shift (2.19 τ) suggests considerable aromaticity. It is reported that this molecule forms weakly coloured charge-transfer complexes with picric acid and trinitrobenzene (Barth & Lawton 1971).

It is still not clear how much of the resonance energy that would have been obtained in the planar form is lost in the shallow-bowl structure. According to SCF–LCAO–MO calculations by Gleicher (1967), even the deep-bowl structure (**9**) retains more than 90% of the π-bonding energy that can be realized in the planar form, and the resonance energy amounts to at least 10 kcal mol^{-1}.

4. Discussion

Having reproduced our old proposal, the outcome is now briefly described. According to Aihara's (1988) topological theory of aromaticity, the resonance energy per π electron (REPE) of **5** is calculated to be 0.0274 β, or about 60% of that of benzene (0.0454 β). It is thus clear that there is no dramatic increase in the conjugative stability in **5**. As far as aromatic stabilization in the football structure is concerned it should be regarded as an extension of two-dimensional aromaticity. Aihara & Hosoya (1988) call it spherical aromaticity.

A true example of 3D delocalization, and consequent stabilization, was later discovered by Schleyer (1987) and described in his award lecture. The superaromatic organic molecule that he presented is a 2π system contained in a tetrahedral space within the adamantane skeleton (**10**). However, how can one create a superaromatic system with more than two electrons? This problem still remains a challenge to chemists.

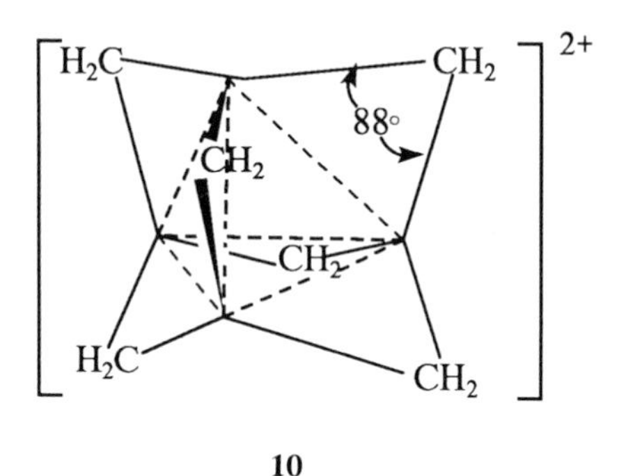

10

It is noteworthy that about the same time as our book was published, two Russian chemists performed Hückel molecular orbital calculation on C_{60} (Bochvar & Galpern 1973). They named it carbo-s-icosahedrene. Their calculations gave the correct picture of the π-orbital system.

It may be of some interest to us to trace how the structure **5** occurred in the minds of these authors. According to their paper, the impetus came from Eaton's (1972) work on the synthesis of peristyrene (**11**), a potential precursor of dodecahedrane, $C_{20}H_{20}$ (**12**). Noting that the C–C–C and C–C–H angles in **12** are expected to be very close to the tetrahedral value, they scanned through the five regular Platonic polyhedrons and 14 semiregular archimedean polyhedrons for similar situations. They found the truncated icosahedron of particular interest, because the carbon atoms in the hydrocarbon analogue ($C_{60}H_{60}$) or carbon analogue (C_{60}) of this polyhedron 'can be considered to be very near to an sp^2-hybridized state'. They thought of dodecahedrane (**12**) and C_{60} (**5**) as the strainless polyhedral molecules consisting exclusively of sp^3- and sp^2-carbon atoms respectively.

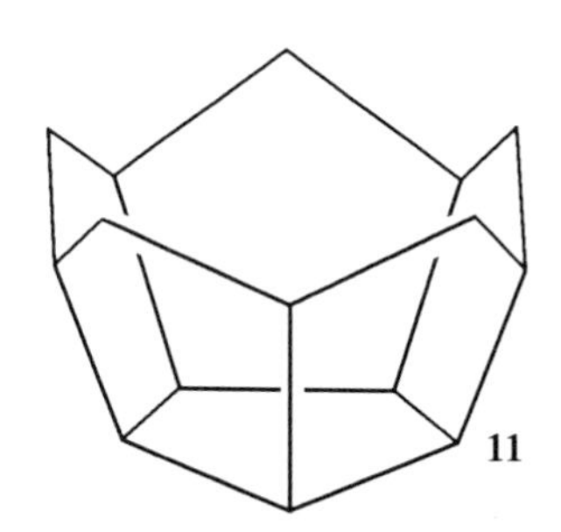

11

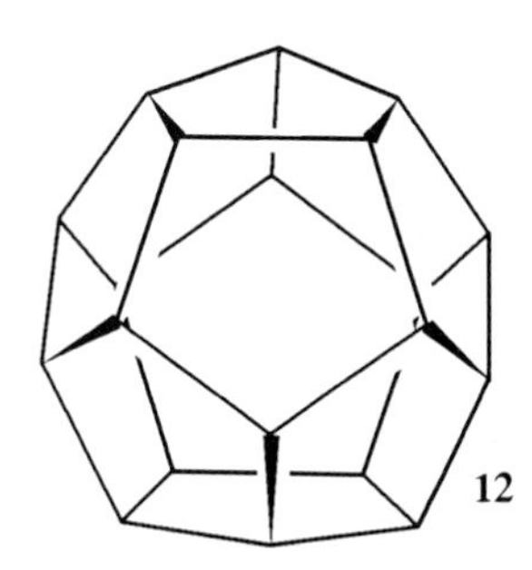

12

The two preliminary studies of **5**, ours and that of Bochvar and Galpern, have different starting points – corannulene and dodecahedrane – but have one thing in common: both parties did not pursue then imaginary molecule any further. Why is it that these early concepts did not develop logically into substantial studies? Perhaps neither of us recognized the hidden potential of the football structure of C_{60} as a new form of carbon and if so we should be categorized as *premature* discoverers, according to Stent (1972, see also Landman 1992).

At least in our own case, however, there were three distinct reasons for not pursuing this problem, and it seems worthwhile to record them here. First, at a time when there was absolutely no anticipation for the possibility that the football structure might form spontaneously, we thought it almost impossible to realize the synthesis of **5**. This fear proved correct: we recently learned that Chapman's arduous attempts to the synthesis in the 1980s were never rewarded (Diederich 1992).

The second reason might sound rather surprising: we actually thought this molecule not particularly interesting (Ōsawa 1991). There are several grounds for this conclusion. As mentioned above, there was no response to our original proposal and we were very discouraged. Furthermore, **5** contains as many as 20 'benzene' rings. Although these rings are deformed out-of-plane, the case of corannulene has taught us that the six-membered rings will tend to remain planar and this should be a source of its stability. Hence, it seemed likely that the stability as well as all other properties of football C_{60} will be very heavily influenced by the presence of so many 'benzenoid' units. Actually, this concern existed from the inception of the football structure. We therefore wished to devise a totally new molecular system of superaromaticity as it would clearly be desirable to have something quite different from the 'benzene analogue'.

The third reason for our premature rejection of **5** was, as has recently been correctly guessed by Karfunkel (1992), the underdeveloped status of computational chemistry in those days: computers large enough to handle C_{60} were not available to us. Even when we performed the first semiempirical MO calculations of C_{60} and C_{70} in 1986, the molecules were rather too large to be studied by routine calculation and we had to work half a year debugging the program until we finally succeeded in detailed vibrational analysis (Slanina *et al.* 1987*a*,*b*, 1989; Rudzinski 1987).

5. Final words

No scientific discovery seems to be totally new, as has been discussed superbly well by Berson (1992) with reference to the discoveries of the Diels–Alder reaction and the Woodward–Hoffmann rule. In the case of C_{60}, the 'near misses' by Iijima (1987) and the unpublished work by Chapman (Diederich 1992), are more pre-eminent examples of the precedence than those described above.

Let us conclude our presentation by a comment on the relative importance between the well-known two steps in the process of discovery: finding and the recognition of finding (Berson 1992). Our inevitable conclusion, after observing such a large number of missed discoveries, is that the latter is much more important and difficult than the former. A finding is usually made by chance, as in the case of the discovery of the C_{60} peak in the mass spectrum of laser-vaporized carbon clusters (Kroto *et al.* 1985). Hence there is not much one can do but to resort to serendipity.

The most crucial moment comes after a finding has been made. The most desirable situation would be that the discoverers themselves recognize the relevance of their finding and explain the relations with the then accepted body of knowledge, using a language that leads others to logically understand the significance of the discoveries. Here a number of novel qualities are required: the imagination to grasp generality on the basis of a small piece of evidence, the talent to give an appropriate name (Nickon & Silversmith 1987) and the ability to communicate well with other scientists. It is truly gratifying to realize that the authors of the 1985 *Nature* paper had all these attributes.

We thank Dr R. N. Compton for a copy of English translation of the Bochvar & Galpern paper.

References

Aihara, J. & Hosoya, H. 1988 Spherical aromaticity of buckminsterfullerene. *Bull. Chem. Soc. Japan* **61**, 2657–2659.

Barth, W. E. & Lawton, R. G. 1966 Dibenzo[ghi,mno]fluoranthene. *J. Am. chem. Soc.* **88**, 380–381.

Barth, W. E. & Lawton, R. G. 1971 The synthesis of corannulene. *J. Am. chem. Soc.* **93**, 1730–1745.

Berson, J. A. 1922 Discoveries missed, discoveries made: creativity, influence, and fame in chemistry. *Tetrahedron* **48**, 3–17.

Bochvar, D. A. & Galpern, E. G. 1973 Hypothetical systems: carbododе-cahedron, s-icosahedron, and carbo-s-icosahedron. *Dokl. Acad. Nauk SSSR* **209**, 610–612.

Diederich, F. & Whetten, R. L. 1992 Beyond C_{60}: the higher fullerenes. *Acc. chem. Res.* **25**, 119–126.

Gleicher, G. J. 1967 Calculations on the corannulene system. *Tetrahedron* **23**, 4257–4263.

Iijima, S. 1987 The 60-carbon cluster has been revealed. *J. phys. Chem.* **91**, 3466–3467.

Karfunkel, H. R. & Dressler, T. 1992 New hypothetical carbon allotropes of remarkable stability estimated by modified neglect of diatomic overlap solid-state self-consistent field computations. *J. Am. chem. Soc.* **114**, 2285–2288.

Krätchmer, W. *et al.* 1990 Solid C_{60}: a new form of carbon. *Nature Lond.* **347**, 354–358.

Kroto, H. W., Heath, J. R., O'Brien, S., Curl, R. F. & Smalley, R. E. 1985 C_{60}: Buckminsterfullerene. *Nature, Lond.* **318**, 162–163.

Landman, O. E. 1992 The fitful path of progress in science: set back by prematurely, spurred by creativity. *The Scientist*, March 30, pp. 12.

Nickon, A. & Silversmith, E. F. 1987 *Organic chemistry: the name game.* New York: Pergamon Press.

Ōsawa, E. 1970 Superaromaticity. *Kagaku* (*Chemistry*) **25**, 854–863. (In Japanese.)

Ōsawa, E. 1991 The origin of C_{60}. *Parity* **6** (11), 58–62. (In Japanese.)

Rudzinski, J. M., Slanina, Z., Togashi, M. *et al.* 1987 Computational study of relative stabilities of C_{60} (I_h) and C_{70} (D_{5h}) gas-phase clusters. *Thermochim. Acta* **125**, 155–162.

Schleyer, P. v. R., Bremer, M., Schötz, K., Kausch, M. & Schindler, M. 1987 Four-center two-electron bonding in a tetrahedral topology. Experimental realization of three-dimensional homoaromaticity in the 1,3-dehydro-5,7-adamantanediyl dication. *Angew. Chem. Int. Ed. Engl.* **26**, 761–763.

Slanina, Z., Rudzinski, J. M. & Ōsawa, E. 1987*a* C_{60} (g), C_{70} (g) saturated carbon vapour and increase of cluster populations with temperature: a combined AM1 quantum chemical and statistical-mechanics study. *Coll. Czech. Chem. Commun.* **52**, 2831–2838.

Slanina, Z., Rudzinski, J. M. & Ōsawa, E. 1987*b* C_{60} (g) and C_{70} (g): a computational study of pressure and temperature dependence of their populations. *Carbon* **25**, 747–750.

Slanina, Z. *et al.* 1989 Quantum-chemically supported vibrational analysis of giant molecules: the C_{60} and C_{70} clusters. *J. Molec. Struct.* (*Theochem*) **202**, 169–176.

Sondheimer, F., Wolovsky, R. & Amiel, Y. 1962 Unsaturated macrocyclic compounds. XXIII. The synthesis of the fully conjugated macrocyclic polyenes cyclooctadecanonaene ([18]annulene), cyclotetracosadodecaene ([24]annulene) and cyclotriacontapentadecaene ([30]annulene). *J. Am. chem. Soc.* **84**, 274–284.

Stent, G. S. 1972 Prematurity and uniqueness in scientific discovery. *Scient. Am.* **227**, 84–93.

Woodward, R. B. & Hoffmann, R. 1970 *The conservation of orbital symmetry.* New York: Academic Press.

Yoshida, Z. & Ōsawa, E. 1971 *Aromaticity.* Kyoto: Kagaku Dojin. (In Japanese.)

Discussion

H. W. Kroto (*University of Sussex, U.K.*). The Hückel rule is $4n+2$. Does this apply to three-dimensional aromaticity? Patrick Fowler has his $6k+60$ rule for C_{60}, but are there different rules for different structures?

P. W. Fowler (*University of Exeter, U.K.*). The only closed shells in fullerenes are those that obey the $6k+60$ rule, and another series that obeys the $30k+70$ rule. If you want closed shells you will not get one, electronically, below C_{60}. So the question of whether C_{60} is aromatic is another story. I think the chemistry is showing that C_{60} is anything but aromatic, and the serendipity was trying to think of an aromatic system, finding one, and then realizing that it wasn't.

E. Wasserman (*Du Pont Experimental Station, U.S.A.*). The word aromatic has so many different meanings, it is very hard to use it in any brief sense, with any degree of accuracy. Certainly, some of the things that we and others have found, on debromination, chlorination and dechlorination of C_{60}, indicates a ready reversion to C_{60}. In planar systems we would regard that as partial evidence for aromaticity. The fact that there are other forms of aromaticity simply means that we are beginning to stretch the term a little far.

P. W. Fowler. You can make a case for the aromaticity of C_{60}. We have called it a molecule of ambiguous aromatic character, because it is clear that it does not follow all of the rules, particularly those set by benzene. When most chemists talk of aromaticity, they probably compare the molecule to benzene as this is the only perfect aromatic compound that we have and it sets a pretty high standard. Consider other compounds that chemists traditionally consider as aromatic, for example, the next highest Hückel molecule, 10-annulene. If we look at the chemistry that has been done on the bridge 10-annulenes, much of it does not follow closely the standards set by benzene. The fact that most derivatives of C_{60} tend to revert to type does make a case for the aromaticity of C_{60}. There are many counter examples, and one is the question of ring currents in C_{60}, which do not appear as simple as in other aromatic compounds. So I do agree that it does not follow all of the rules. Adding to the example that you gave of corannulene (**6**) in which the five-membered ring at the centre of the molecule has six π-electrons, and the outer perimeter had 14 π-electrons, Harry Scott has just succeeded, with Mordicai Rabinowitz, in adding four electrons to corannulene. You now have six electrons in the centre of the molecule, and 18 electrons on the periphery. In that way it can be imagined as an aromatic outer 18 π-electron system, and an inner six π-electron system, which nicely confirms the things you were talking about.

E. Ōsawa. I have something to add on the tendency of chlorine and bromine derivatives of C_{60} to revert to C_{60}. One of the reasons for this tendency is not only aromaticity, but also the strange properties of the σ-bonds extending outside the structure. These σ-electrons have π-electrons as neighbours; a unique situation in organic chemistry. As a result of σ/π interactions, the σ-bond will be very weak, so this could be one of the reasons for the reversal of halogenation.

Dreams in a charcoal fire: predictions about giant fullerenes and graphite nanotubes

By David E. H. Jones
Physical Chemistry Department, University of Newcastle upon Tyne, NE1 7RU, U.K.

The early prediction of hollow graphite molecules suggested that they should be supercritical under ambient conditions. This is not true of C_{60}, but might still be true of higher fullerenes and graphite nanotubes of large diameter.

---

1. Introduction

My title refers to the celebrated vision of Kekulé, one of the founders of the concept of chemical structure. In 1865, staring drowsily one evening into the fire, he saw in a dream the cyclic structure for benzene, that fundamental unit of all aromatic molecules, and of graphite and the fullerenes. In his reverie, he imagined the atoms gambolling before his eyes…'one of the snakes had seized hold of its own tail, and the form whirled mockingly before my eyes' (Kekulé 1890). In this paper I deal, not so much with the recent triumphs of the identification and bulk preparation of buckminsterfullerene, as with its imaginative prehistory. This begins with Dalton's atomic theory, elaborated from 1803 onwards. Despite a very promising start, atomic theory languished for decades as merely a sort of useful metaphor. One good reason was its failure to come up with consistent atomic weights for the elements and formulae for their compounds. Whether, for example, the atomic weight of oxygen was 8 and water was HO, or whether it was 16 with water as H_2O, remained uncertain for half a century.

And yet shortly after Dalton proposed his theory, the whole problem had been solved (Avogadro 1811). On the assumption that equal volumes of gas under the same conditions of temperature and pressure contained the same number of molecules, Avogadro was able to allocate consistent formulae to the known gases and vapours, and consistent atomic weights to the elements composing them. The one novel assumption he had to make was that the common elemental gases, hydrogen, oxygen, nitrogen and the like, were composed of double atoms: H_2, O_2, N_2 and so on.

Avogadro's solution seemed deeply repugnant to the chemists of the day; so repugnant that it was not so much argued against as simply ignored. At this distance in time, it is hard to see why. My guess is that monoelemental compounds like O_2 seemed completely incredible. All chemical experience, buttressed by the amazing successes of electrochemistry, suggested that elements combine because of some difference of polarity; the more dissimilar two elements were, the more likely they would be to combine. So how could two identical atoms react together? As a result of this mind-set, chemistry remained in needless confusion from 1811 till 1860, a whole working lifetime. Avogadro's hypothesis was finally accepted only after a special exposition of it (Cannizzaro 1858) was distributed at the con-

Phil. Trans. R. Soc. Lond. A (1993) **343**, 9–18

Printed in Great Britain

ference of Karlsruhe in 1860. So I would like to propose Avogadro, the inventor of the whole concept of monoelemental compounds, as the ultimate godfather of buckminsterfullerene. His is the original insight which underlies our familiar acceptance of, for example, O_2, P_4, S_8, B_{12}, and now the latest and most striking example, C_{60}.

2. Early predictions of hollow graphite molecules

Carbon, of course, is the ideal element for simple-minded chemists. Obey a few elementary valency rules, and almost any organic structure you can doodle would exist, if you could make it. And this is perhaps the justification for my own simple-minded contribution to the story of C_{60}. For many years I have maintained a scientific *alter ego*, Daedalus, whose musings used to appear in *New Scientist* but now appear in *Nature*. Daedalus launches scientific proposals which are intended to fall in that uneasy no-man's-land between the clearly feasible and the clearly fantastic. His aim is inevitably rather erratic, and many of the attempts land on one side or the other. An account of some of Daedalus's chemical proposals has appeared in *Chemistry in Britain* (Jones 1987). His greatest moment came late in 1966, when he proposed the hollow-shell graphite molecule (Jones 1966).

All these years later, I can't remember just what triggered that particular proposal. But once I had the idea, it seemed to me that the physical properties of such structures would be highly unusual. I was thinking of what would now be called the giant fullerenes. These big, light molecules should be mainly empty space, and so would have a very low bulk density. In addition, I felt that they would interact with each other very weakly indeed. Despite their enormous molecular weight, they might be extremely volatile. After some naive calculations, I predicted for my hollow graphite molecules the properties below.

1. Bulk density intermediate between those of liquid and gas (e.g. 40 kg m^{-3}).
2. Supercritical at ambient pressure and temperature.
3. Pourable as light, loose liquids with an ill-defined surface.
4. Able to trap smaller molecules in their hollow interiors, and possibly to exchange them.
5. Able to act as lubricants, with the molecules rolling like ball-bearings.

Of these predictions, 5 was pure provocation: I didn't believe it at the time, and indeed it is clearly absurd. At the molecular level, lubrication functions by sliding, not rolling. No molecule, certainly not a hollow one, could act as a frictionless undeformed rolling element, and in any case would need a molecularly smooth track on which to do so. The other predictions seemed to me about right for Daedalus: challenging, but not obviously beyond belief.

As with most of Daedalus's output, this suggestion met with no reaction from the scientific community. But some years later I published a book containing a number of Daedalus's better schemes (Jones 1982). Many of the articles were enhanced with diagrams and additional background material which had not been included in the original publications. Among these extended items was the article on the hollow graphite molecule. I have to salute the forbearance with which the editor, Michael Rodgers, tolerated the heavy technical additions to what was intended as basically a popular and light-hearted book. A sterner editor would have put a blue pencil through the lot: when my position among the originators of the hollow graphite

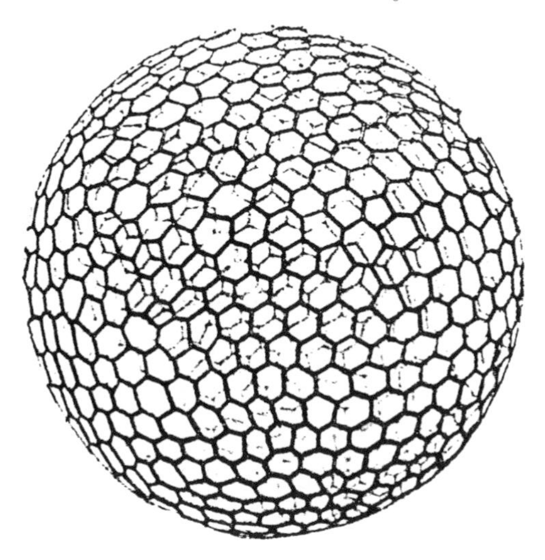

Figure 1. The diatom *Aulonia hexagona* is an almost perfect 100000-fold model of a giant fullerene with about 1200 carbon atoms. (From *On growth and form*, by W. D'Arcy Thompson. Courtesy of Cambridge University Press.)

molecule would have been far weaker. In the event, he allowed me to include a detailed discussion of these hypothetical molecules, including a biological analogy (figure 1) and some comments on the theory of polyhedra, in particular the rather surprising fact that any hexagonal mesh needs exactly 12 pentagons to make it into a closed shell; a rule discovered by Euler, but which I found in that wonderful book *On growth and form* (D'Arcy Thompson 1942). I was particularly concerned to find the largest size that hollow graphitic molecules could be made before they would collapse by inversion. A highly suspect calculation suggested that the biggest stable one would be C_{260000}, or thereabouts. It didn't occur to me to wonder how small they could be made; a sad and crucial failure of chemist's intuition!

3. Buckminsterfullerene arrives

In 1985 the remarkable predominance of the C_{60} fragment in the mass spectra of laser-ablated graphite was noticed, and the hollow buckminsterfullerene structure was suggested (Kroto *et al.* 1985). This brilliant proposal owed nothing to my musings in this area; in fact Harry Kroto only learnt of them later through Martyn Poliakoff at Nottingham University, who had a copy of the book. The only useful contribution which it made to the fast-developing saga was in pointing out the necessity for 12 pentagons in any closed-shell hexagonal structure: the C_{60} structure had been proposed without guidance from Euler's rule. None the less, Harry had the courtesy to telephone me with congratulations on a remarkable prediction. One of Daedalus's notions had turned out surprisingly close to possible reality.

Between 1985 and 1990, the structure and bulk properties of buckminsterfullerene remained uncertain. I was able to maintain an optimistic hope that predictions 2–4 would turn out to apply even to this small and compact molecule (prediction 1, a consequence of the empty space inside a large hollow molecule, could only be true of giant fullerenes). But when C_{60} was produced in macroscopic quantities (Krätschmer

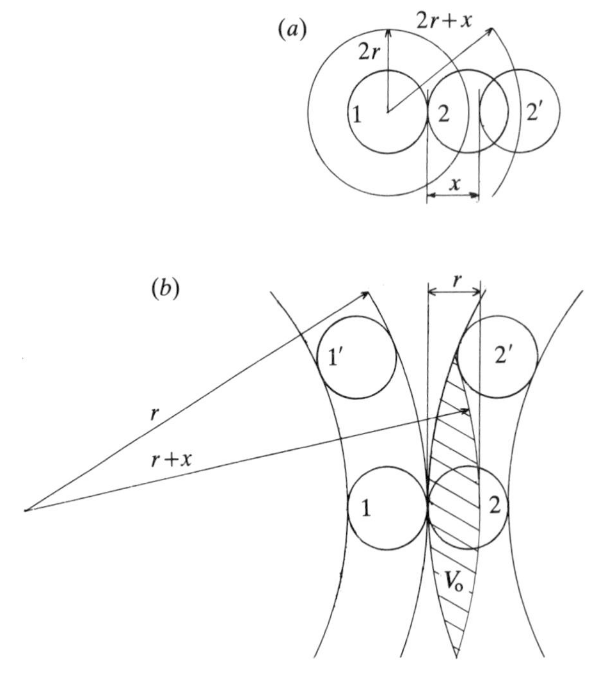

Figure 2. (*a*) Interaction of small spherical molecules. The molecule at 2, on the inner edge of a square potential well of width x, is 'in contact' with the central molecule 1. At 2′ it is on the outer edge of the well. The radius $2r$ defines the exclusion volume around the central molecule; the shell between radius $2r$ and $(2r+x)$ defines its interaction volume. (*b*) Interaction of spherical shell-molecules. Atoms 1 and 2 on their respective shells are 'in contact' on the inner edge of a square well of width x. Atoms 1′ and 2′ are beyond the outer edge of the well. If the shell-atoms are taken as uniformly 'smeared' around the shells, the energy of interaction between the molecules should be approximately proportional to the 'overlap volume' V_0, the region in which the shells are closer than x.

et al. 1990), it became clear how wrong I was. So far from being supercritical at ambient conditions, or even remarkably volatile for its molecular weight, C_{60} does not melt, and is so involatile that its variation of vapour pressure with temperature has to be measured by mass spectroscopy (Pan *et al.* 1991). Only prediction 4, the ability of a hollow graphite molecule to entrap small atoms or molecules inside it, has turned out correct (Heath *et al.* 1985; Weiss *et al.* 1988).

The determined theorist should make an orderly retreat in the face of the advancing facts. Giant fullerenes have not yet been made in bulk, so I am still free to maintain my original claims in respect of these as yet hypothetical substances. Predictions 2 and 3 were based essentially on my intuitions about the weakness of the expected interaction between big spherical molecules. I buttressed them with some highly over-simplified calculations, not hitherto published. They are more of a speculation than a rigorous argument.

4. Properties of giant fullerenes

Prediction 3 is a consequence of high molar mass. The 'scale height' of a gas in a gravitational field is that height for which its pressure decreases by 1/e. In the perfect-gas approximation, it is given by $h = RT/gm$, where m is the molar mass of the gas. For air at ambient temperature, $h = 8.8$ km. If my maximally sized giant fullerene C_{260000} were a perfect gas, it should have $h = 8$ cm. Supercritical C_{260000} would be a very imperfect gas. It would therefore have $h < 8$ cm, and could be retained safely in a fairly tall open laboratory beaker.

Prediction 2, that such big molecules ought to be supercritical gases under ambient conditions, is supported by a more contentious argument based on their equation of state. The liquid state, and condensed phases in general, exist because of the short-range attractive forces between molecules. The simplest of all equations of state, Boyle's Law, $PV = RT$, makes no provision for attractive forces and does not predict condensed phases. The next approximation is Van der Waals' equation:

$$(P+a/V^2)(V-b) = RT.$$

This equation acknowledges that real molecules have size. They have an exclusion volume, defined as the region around the molecule from which the centre of any other molecule is excluded. This is allowed for by the constant b, which is usually taken as equal to half the molar exclusion volume. The equation also recognizes the existence of a sphere of influence around each molecule, an interaction volume within which any other molecule will experience a force of attraction. This force is usually represented by a Lennard–Jones 6–12 potential. The derivation below follows a simpler treatment (Flowers & Mendoza 1970) in which the potential is taken as a square-well function as deep as the Lennard–Jones minimum (figure 2*a*). Its width x is chosen to give the same volume-integral, and defines an 'interaction volume' V_x around the molecule, which will contain the centre of any molecule in the square well. This form of molecular pair potential then appears in the Van der Waals equation as the constant a, equal to half the product of the molar interaction volume and the molar interaction energy.

The Van der Waals equation is one of those happy approximations which somehow make reasonable predictions well outside the region in which their assumptions are valid. It can even be used to predict the critical constants of a gas. A Van der Waals gas should have a critical pressure and temperature given respectively by

$$P_{c,\mathrm{vw}} = a/27b^2 \quad \text{and} \quad T_{c,\mathrm{vw}} = 8a/27Rb.$$

Reasonable choices of a and b can often predict the critical constants of real gases to within 10 or 20% of their observed values.

An estimate of the critical constants of a substance composed of large hollow-shell molecules can therefore be obtained from its values of a and b. Consider two such molecules in contact (figure 2*b*). An atom in one shell finding itself close to an atom in the other, will experience an attractive force; pairs of atoms further apart will feel a much smaller force and will contribute little to the interaction. If each shell is taken as a uniform distribution of atoms exerting a square-well potential, the energy of interaction reduces to the product of the energy-density of the potential well and a lenticular volume of overlap V_o. The central width of this volume is x, the square-well width. For r much greater than x, V_o (and therefore the energy of interaction) increases linearly with r. The interaction volume V_x is defined as in figure 2*a*: a shell of radius $2r$ and thickness x. For r much greater than x, it equals $16\pi r^2 x$ and therefore increases as r^2. The constant a is half the product of the molar interaction energy and the molar interaction volume. So it increases as $r^2 r$, i.e. as r^3. The constant b, proportional to the exclusion volume of the shell-molecule, also increases as r^3.

Thus for large shell-molecules the factor a/b, which determines the Van der Waals critical temperature $T_{c,vw}$, should be constant with change of radius. Extending the treatment to the case of figure 2a in which r is comparable with x (in fact $r = 0.922x$, after Flowers & Mendoza), and making the simplifying assumption that x is effectively unchanged, gives this constant as about 0.45 times the value of a/b for the equivalent small molecule. By contrast, a/b^2, which determines the critical pressure $P_{c,vw}$, should decline as r^3. Accordingly, a hollow molecule of reasonable size, say five times the radius of a simple fully dense molecule, should have a T_c of about half the normal value, and a P_c about 1% of it. T_c and P_c for common simple fluids seldom exceed 500 K and 70 atm, so reductions of this magnitude might well bring them down to ambient temperature and pressure or below. Hence a substance consisting of hollow spherical molecules could quite plausibly exist as a supercritical fluid under ordinary conditions. The low critical pressure seems a more secure prediction than the low critical temperature.

It is worth noting that this calculation pays no attention to the mass of the molecule. The liquid or gaseous state of the material, and its critical constants, depend only on the molecular radius and pair potential. This runs counter to our chemical intuition; it is natural to imagine that big molecules form involatile solids or high-boiling liquids because they are too heavy to take off. This cannot be true. Big molecules have low volatility because they interact strongly with their neighbours. Even a heavy molecule like UF_6 can be surprisingly volatile; its spherical symmetry and weak pair-potential reduces its interaction with its neighbours. Molecular mass must have some restraining influence on volatility; after all, critical temperature and pressure increase slightly along the series H_2, D_2, T_2. The low densities of hollow-shell molecules like the giant fullerenes should at any rate make them more volatile than fully packed molecules of equivalent radius.

The above calculation is clearly too impressionistic to support precise predictions. The case of the giant fullerenes can be studied more carefully in the light of a recent calculation of a 6–12 potential function for C_{60} molecules based on values for graphite (Girifalco 1992). Contrary to my expectations, the C_{60} molecule interacts very strongly with its nearest neighbours: its second virial coefficient $B(T)$ is two orders of magnitude greater than for the common gases. A Van der Waals gas has $B(T) = b - a/RT$. Values of a and b for C_{60} can be derived by fitting an equivalent square well to Girifalco's potential function, but they do not reproduce his $B(T)$. The best agreement is obtained from a square well with an outer limit rather beyond that suggested by the potential function. The resulting values of a and b then predict $T_c = 1300$ K at $P_c = 12$ atmospheres, and fit $B(T)$ in the neighbourhood of this T_c. These critical constants do not seem unreasonable for C_{60}. A fullerene with five times the radius, say C_{1500}, should then on the argument above have $T_c = 650$ K at $P_c = 0.1$ atmospheres. The high critical temperature is a little disappointing. A fluorinated or hydrogenated giant fullerene, whose molecules would attract each other less strongly, should have a lower one.

Despite the tenuity of the argument, I continue to feel that large enough hollow spherical molecules should show very low critical constants, or at any rate a very low critical pressure. But even if fullerene molecules large enough to test this claim are ever obtained in bulk, they could still escape on a technicality. Molecular models (Kroto & Mackay 1988) show them not to be spherical, but distinctly icosahedral. Their mean interaction energy, averaged over all orientations of molecular collision, might not be the same as that of spheres of an equivalent diameter; though the difference should not be great. In any case, such molecules when synthesized may well turn out to be spherical or nearly so, being inflated by trapped internal gas like tiny footballs.

5. Predictions of giant graphite nanotubes

Shortly after the mass-spectroscopic identification of the C_{60} fragment, and the proposal of its spherical structure, Daedalus attempted to maintain his lead in the field by suggesting the synthesis of what would now be called giant graphite nanotubes (Jones 1986). He proposed to make these molecules by a mechanism which has since been disproved for C_{60} (O'Keefe *et al.* 1986; McElvany *et al.* 1987), the laser-detachment from a graphite surface of single graphite sheets, which subsequently curl up. A laser interference-pattern of closely spaced lines, he claimed, would detach long thin graphitic strips; these would roll or curl up into cylinders. Again I was intrigued by the possible volatility and even supercriticality of these large, empty, cylindrical molecules. A calculation like that of §4 should apply to them as well. I further suspected that, like high polymers in liquid solution, such long volatile molecules would make the air extremely viscous, even if present in very low concentrations. They could thus be used as a blanketing agent to slow cars on motorways, as a sound-damping medium, and as a shield against bullets and missiles.

Graphite nanotubes have since been identified by electron microscopy of the products of carbon-arc synthesis (Iijima 1991); another successful prediction for Daedalus. But the bulk synthesis of giant nanotubes, like that of giant fullerenes, is still beyond us. Such large single-sheet graphite cylinders (they should perhaps be called microtubes, or even millitubes) might possibly be grown, or even blown. Graphite is made commercially by heating carbonized precursor material to 3000 °C. At high-temperatures dislocations in a graphite sheet become quite mobile (Scott & Roelofs 1987). The small disordered sheets of the starting material can grow and seek the thermodynamic advantages of large-scale order (Maire & Mering 1960). They can be oriented and extended by simple stretching; oriented graphite fibres are made by the controlled simultaneous stretching and graphitization of a polymeric precursor fibre. So Daedalus imagines (Jones 1991) that at a high enough temperature, a graphite sheet approaches a condition something like a liquid soap-film. It could be stretched in two dimensions, so that a giant fullerene molecule might even be expanded like a soap-bubble. With still more help from some as yet unimagined natural micro-phenomenon, a giant nanotube could perhaps be cylindrically extruded and inflated like blown polyethylene film. Such a tube would have to be rapidly quenched, for a cylindrical liquid film becomes unstable if its length exceeds π times its diameter. It breaks up into a chain of bubbles, which in this case would be giant fullerene molecules.

At this stage, however, my argument has degenerated from defensible speculation into a pure dream in a charcoal fire. I look forward to the eventual bulk production, by unguessed chemistry, of giant carbon fullerenes and nanotubes. Even if they do not have the properties I originally imagined for them, I am sure they will be fascinating substances.

References

Avogadro, A. 1811 Essay on a manner of determining the relative masses of the elementary molecules of bodies, and the proportions in which they enter into their compounds. *J. de Physique* **73**, 58–76. (English edn: *Dalton, J. and others: Foundations of the molecular theory*. Edinburgh: William Clay, Alembic Club Reprint no. 4 (1893).)

Cannizzaro, S. 1858 *Il Nuovo Cimento* **7**, 321–366. (English edn: *A sketch of a system of chemical philosophy*. Edinburgh: William Clay, Alembic Club Reprint no. 18 (1911).)

Flowers, B. H. & Mendoza, E. 1970 *Properties of matter*, pp. 188–210. London: John Wiley and Sons.

Girifalco, L. A. 1992 Molecular properties of C_{60} in the gas and solid phases. *J. phys. Chem.* **96**, 858–861.

Heath, J. R., O'Brien, S. C., Zhang, Q., Liu, Y., Curl, R. F., Kroto, H. W., Tittel, F. K. & Smalley, R. E. 1985 Lanthanum complexes of spheroidal carbon shells. *J. Am. chem. Soc.* **107**, 7779.

Iijima, S. 1991 Helical microtubules of graphitic carbon. *Nature, Lond.* **354**, 56.

Jones, D. E. H. 1966 *New Scientist* **35** (519), 245.

Jones, D. E. H. 1982 Hollow molecules. In *The inventions of Daedalus*, pp. 118–119. Oxford: W. H. Freeman.

Jones, D. E. H. 1986 *New Scientist* **110** (1505), 88 and (1506), 80.

Jones, D. E. H. 1987 The chemistry of Daedalus. *Chem. Brit.* **23**, 465–468.

Jones, D. E. H. 1991 Ghostly graphite. *Nature, Lond.* **351**, 526.

Kekulé, A. 1890 *Berischte der Deutschen Chemischen Gesellschaft* **23**, 1302. (transl. O. T. Benfey, 1958 *J. Chem. Educ.*, **35** (1) 21–23).

Krätschmer, W., Lamb, L. D., Fostiropoulos, K. & Huffman, D. R. 1990 Solid C_{60}: a new form of carbon. *Nature, Lond.* **347**, 354.

Kroto, H. W., Heath, J. R., O'Brien S. C., Curl, R. F. & Smalley, R. E. 1985 C_{60}: Buckminsterfullerene. *Nature, Lond.* **318**, 162.

Kroto, H. W. & McKay, K. 1988 The formation of quasi-icosahedral spiral shell carbon particles. *Nature, Lond.* **331**, 328.

Maire, J. & Mering, J. 1960 Croissance des Dimensions des Domaines Cristallins au Cours de la Graphitation du Carbone. In *Proc. Fourth Conf. on Carbon*, p. 345. New York: Pergamon Press.

McElvany, S., Nelson, H. H., Baronavski, A. P., Watson, C. H. & Eyler, J. R. 1987 FTMS studies of mass-selected large cluster ions produced by direct laser vaporization. *Chem. Phys. Lett.* **134**, 214–219.

O'Keefe, A. O., Ross, M. M. & Baronavski, A. P. 1986 Production of large carbon cluster ions by laser vaporisation. *Chem. Phys. Lett.* **130**, 18–19.

Pan, C., Sampson, M. P., Chai, Y., Hauge, R. H. & Margrave, J. L. 1991 The heats of sublimation from a polycrystalline mixture of C_{60} and C_{70}. *J. phys. Chem.* **95**, 2944–2946.

Scott, L. T. & Roelofs, N. H. 1987 Benzene ring contractions at high temperatures. Evidence from the thermal interconversions of aceanthrylene, acephenanthrylene and fluoranthrene. *J. Am. chem. Soc.* **109**, 5461–5465.

Thompson, D'A. W. 1942 *On growth and form*, pp. 708, 738. Cambridge University Press.

Weiss, F. D., O'Brien, S. C., Elkind, J. L., Curl, R. F. & Smalley, R. E. 1988 Photophysics of metal complexes of spheroidal carbon shells. *J. Am. chem. Soc.* **110**, 4464.

Discussion

E. Wasserman (*The DuPont Company, U.S.A.*). When you talk about the possible phases of something like C_{60} (is it a gas, a liquid or a low density liquid of high compressibility), we really have to compare it with another phase which may be accessible under the same temperature and pressure conditions. In many such cases, some of the features of C_{60} are due to intermolecular interactions, in some of the more condensed phases, rather than to individual molecular properties that you were concentrating on. For example, the very strong tenacity of one C_{60} molecule to bond to another, as well as to incorporate solvent molecules in the interstitial spaces, depends critically on how well they seem to fit together, as well as to the intrinsic forces that may be found in smaller molecules. We find that if you have small degrees of substitution of C_{60}, for example, alkyl groups, the volatility increases dramatically.

We see mass spectra at much lower temperatures than we do for C_{60} itself. So, if the packings of the condensed planes are interfered with, you may be able to increase volatility, and go towards some of the directions you were referring to.

D. E. H. Jones. My original feeling was that the individual graphite planes have very little feeling for one another, and that is why graphite is a lubricant. I had guessed the interaction between graphite sheets was quite low. Clearly, if that is true, it does not transfer to the fullerene structures, which is a surprise, but I am glad that there is a way of getting around it by suitable substitution.

E. Wasserman. I think one of the reasons that graphite is a lubricant, is that as one sheet moves relative to another, you don't lose interactions. The key point is the ease of rotation of C_{60} in the solid.

E. J. Applewhite (*Washington, U.S.A.*). I enjoyed your comments on the giant fullerenes, which reminded me of one of Buckminster Fuller's most way out ideas. Please don't judge Fuller on this particular idea, but I will tell you about it in the spirit of interdisciplinary interaction. If geodesic balls are built big enough, the weight of the construction materials are minuscule, compared to the weight of air contained within the structure. If the sun heated the interior of the dome, then the weight of the air inside the dome, plus the weight of the structure, would be less than the weight of air normally, and the structure would float. Control of the heating procedure would control the altitude of the bucky-ball.

D. E. H. Jones. A nice notion; a flying greenhouse!

H. W. Kroto (*University of Sussex, U.K.*). When we attempted to extract C_{60} we were not sure of the state, solid, even a gas. It might have been volatile, because UF_6 is volatile, and C_{60} is like a rare gas in many ways. Radon is only half the weight, so we thought that if we were producing it, it might have evaporated before you could 'catch it'. So there are interesting aspects of volatility, especially where C_{60} is concerned. The second point is that at ICI, Margaret Steel has produced something similar to your 'pipe'. I believe her group used a CO_2 laser on a spinning graphite cylinder, and spun off some small balls. They were not hollow, but this convinced me that these were ways of making the concentric shell structures that Iijima observed for the first time many years ago. I am unhappy that we were not able to collaborate further with that group at the time, because the evidence she had was that they were concentric shell structures. She was interested in producing carbon fibres, and spinning these things, but they were very close.

S. Iijima (*NEC Corporation, Japan*). The giant fullerene structures are not as you mentioned. Already I, and other groups, have done similar experiments that involved having the metal in a small cluster, for example nickel or iron, and carbonising the metal. I have made thin graphitical films all around the metal particle, so it is similar to the structure you have mentioned.

R. C. Haddon (*AT & T Bell Laboratories, U.S.A.*). The question of what holds C_{60} together, and just how strong the forces are, is a very interesting one. Work we have done recently has shown that it is possible to make membranes of C_{60}, i.e. a free standing film of C_{60} molecules. It is not obvious that this is possible, because you are asking a sheet of C_{60}, a few thousand ångstroms thick, to support itself in a free standing arrangement, which is composed of C_{60} molecules, each of which is spinning in space millions of times a second. So I am particularly interested in learning about the forces that hold C_{60} together, and whether or not they differ from those that hold other Van der Waals's solids together.

D. E. H. JONES. I would not expect C_{60} to have any tensile strength at all. It is intriguing and remarkable. This phenomena is not seen with other spinning solids, such as camphor.

C. T. PILLINGER (*Open University, U.K.*). People have talked about inflating C_{60}, but what happens when you compress? I have heard reports that diamonds are produced. What are your thoughts on the subject?

D. E. H. JONES. In my original suggestion, I was worried that the structure would collapse in on itself, and that you would gain a lot of surface energy of interaction. I did some rather unbelievable calculations to see how big it would be, before it would spontaneously invert. My feeling is that a structure enclosing a large empty space will be unstable, and that it will not tolerate compression. That again applies to the notion of the giant fullerenes. C_{60} itself looks small enough and neat enough to withstand any such deformation, but one might expect the IR spectra of the larger fullerenes to show some worrying modes, before the thing collapses.

On the formation of the fullerenes

By R. F. Curl
Chemistry Department and Rice Quantum Institute, Rice University, Houston, Texas 77251, U.S.A.

The chemistry by which the closed-cage carbon clusters, C_{60} and C_{70}, can be formed in high yield out of the chaos of condensing carbon vapour is considered. Several mechanisms for this process that have been proposed are critically discussed. The two most attractive are the 'pentagon road' where open sheets grow following the alternating pentagon rule and the 'fullerene road' where smaller fullerenes grow in small steps in a process which finds the buckminsterfullerene (C_{60}) local deep energy minimum and to a lesser extent the C_{70} (D_{5h}) minimum. A clear choice between the two does not seem possible with available information.

1. Introduction

The observation (Kroto *et al.* 1985) that the truncated icosahedron molecule, C_{60}^{BF} (buckminsterfullerene), is formed spontaneously in condensing carbon vapour was greeted by some in the chemical community with some doubt. It seemed incredible that this highly symmetrical, closed, low entropy molecule was forming spontaneously out of the chaos of condensing high-temperature carbon vapour. We still believe that the formation of C_{60}^{BF} in supersonic cluster beam sources must be a relatively minor channel, probably accounting for less than 1% of the total carbon. Thus when C_{60}^{BF} was finally isolated from graphitic soot (Krätschmer *et al.* 1990), it came as a surprise that the C_{60}^{BF} plus C_{70} yields were as large as 5%. Later yields have improved substantially, for example Parker *et al.* (1991) obtained a total yield of C_{60}^{BF} of about 20% with total extractable fullerene yields totalling 44% from a carbon arc soot. Thus conditions can be found where C_{60}^{BF} and fullerene formation in carbon condensation can hardly be called a minor channel.

Several proposals have been put forward to account for the formation of C_{60}^{BF} and the other fullerenes. Before reviewing these proposals, it seems appropriate to consider the key observations about fullerene synthesis, the kinds of carbon clusters possible, their energetics, and the thermodynamic and kinetics of carbon clustering.

2. Key observations on fullerene synthesis

(a) *Effect of buffer pressure in arc synthesis*

Krätschmer *et al.* (1990) vaporized graphite in the presence of He buffer gas, examined the soot collected and observed that the intensity of the UV and infrared absorption features characteristic of C_{60} were buffer gas pressure dependent. At a pressure of about 1.3×10^3 Pa (10 Torr), these features are absent, but are clearly

Phil Trans. R. Soc. Lond. A (1993) **343**, 19–32
Printed in Great Britain

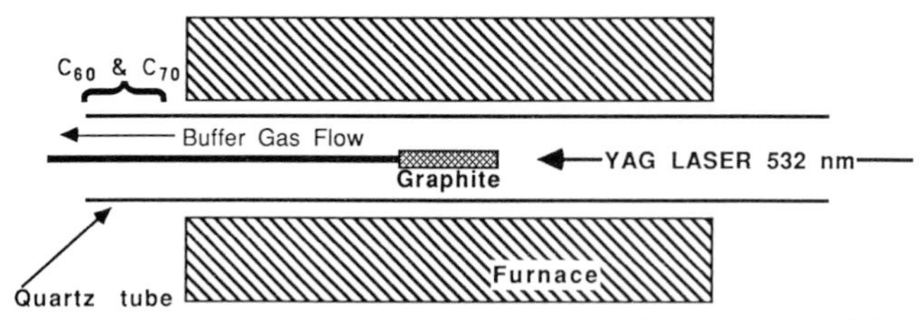

Figure 1. Tube furnace laser vaporization apparatus used by Haufler *et al.* (1991) to produce high yields of C_{60}^{BF}. The 532 nm doubled Nd: YAG pulse vaporizes C from the rotating graphite target into the inert carrier gas stream. The fullerenes condense just outside the oven on the tube wall.

present when the buffer gas pressure raised to 1.3×10^4 Pa (100 Torr). Extensive subsequent experiments by many groups indicate that C_{60}^{BF} yields depend only slightly on the buffer gas pressure between 10^4 and 5×10^4 Pa (100–400 Torr), but decrease rapidly at lower or higher buffer gas pressures.

(b) Effect of wall temperature in laser vaporization

Haufler *et al.* (1991) set up the apparatus shown in figure 1 to explore the effect of wall temperature on fullerene yields by laser vaporization. In this apparatus graphite is vaporized from the end of the graphite rod by a softly focused pulsed visible laser (Nd: YAG, 532 nm, 5-ns pulse, 300 mJ pulse^{-1}). When the walls of the tube are at room temperature no fullerenes are found in the resulting soot; only when the walls of the tube were heated to 1000 °C were any fullerenes obtained. As the furnace temperature was increased further, the yields increased reaching 20% C_{60}^{BF} at 1200 °C, which was the maximum temperature obtainable with this apparatus.

The observations by Krätschmer *et al.* and Haufler *et al.* can be interpreted as indicating that the processes forming C_{60}^{BF} require elevated temperatures for some time, i.e. involve reactions with substantial activation energies. In the Krätschmer *et al.* experiments, the higher buffer gas pressure confines the carbon vapour to a high temperature region near the heat source. In the laser vaporization experiments, a high temperature is supplied directly.

3. Energetics of carbon clustering

The structure and stability of carbon clusters has been of intense interest since the pioneering work of Pitzer & Clementi (1959). The general picture which has emerged is that for the smaller C_n clusters ($n < 10$) that linear structures generally are energetically most favoured. At about C_{10} these linear structures close into rings which presumably persist as the lowest energy form to about C_{20}. In the region C_{20} to C_{30} (the 'forbidden zone') it is unclear which structures are energetically most favourable. Above about C_{30}, the lowest energy structures up to several hundred carbon atoms appear to be the fullerenes.

Our focus is upon the fullerenes. In particular, the stability of C_{60} and C_{70} relative to the other fullerenes and to each other is of primary interest because these two species are the fullerenes produced in highest yield in carbon arcs and in combustion. Quite a few calculations mainly using various semi-empirical theories have been carried out to determine the stability of the fullerenes. Figure 2 shows the recent results of Scuseria's group (Strout *et al.* 1993), who carried out minimal basis set STO-3G SCF calculations on several fullerenes. The smooth curve fitted through

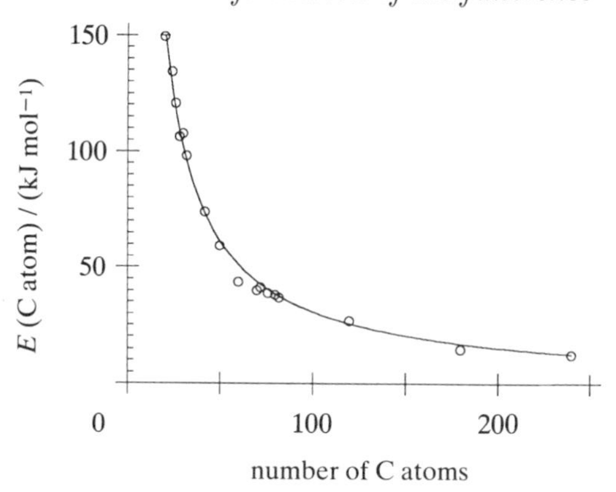

Figure 2. STO-3G restricted Hartree–Fock SCF energies per C atom of representative fullerenes relative to that calculated for single sheet graphite from the work of Strout *et al.* (1992). $E/\text{atom} = 3071/N_c$.

these energies shows that C_{60} and to a lesser extent C_{70} are, as expected, unusually stable. Note particularly that C_{70} is energetically more stable on a per carbon atom basis than C_{60}.

4. Thermodynamics of carbon clustering

A model for fullerene formation based on thermodynamic considerations would be much simpler than one based on kinetics. Fullerene formation might be modelled by supposing that small species are initially produced in a carbon arc at equilibrium with the electrode surface and that this carbon vapour diffuses through the inert gas into a cooler region where the equilibrium vapour density is much less than the actual density. There the fullerenes could be formed at thermodynamic equilibrium assuming that nucleation of graphite is kinetically forbidden. Then this equilibrium distribution could be quenched upon further diffusion into a cooler region. This model is highly attractive because it avoids any consideration of complex kinetics; unfortunately, it is demonstrably incorrect.

Slanina *et al.* (1989) have considered the thermodynamics of the fullerenes in carbon vapour, and their work demonstrates that the C_{70}/C_{60} ratio of approximately unity (ratios obtained vary roughly from 0.1 to 10) found in extracts from graphitic soot is incompatible with thermodynamic equilibrium being established between these two species when they were formed. In their work, the sum of the partial pressures of C_{60} and C_{70}, $\bar{P}_{60\text{–}70}$, at which the species are present in equal concentrations at equilibrium was calculated, and $\bar{P}_{60\text{–}70}$ is never larger than 10^{-13} bar below 5000 K. The observed C_{60} and C_{70} yields cannot be reached with such a small partial pressure of product. If the sum of the C_{60} and C_{70} partial pressures is raised, C_{70} rapidly becomes totally dominant at thermodynamic equilibrium.

5. Energetic effects and C_{60} isomers

Energetic and thermodynamic considerations can play a pivotal role in fullerene formation by providing a driving force for selecting C_{60}^{BF} and providing a large activation barrier to further reactions of C_{60}^{BF}. The potential surface minimum corresponding to buckminsterfullerene is very low in comparison with that of the other fullerene isomers of C_{60} (which in turn almost certainly have lower energies than non-fullerene C_{60} isomers). Further, the energy per carbon atom of C_{60}^{BF} is much lower than any isomer of C_{58} and C_{62} as is indicated in figure 2. Locally (in terms of carbon cluster size) C_{60}^{BF} is a deep potential well.

In addition, C_{60}^{BF} has no weak point for chemical attack, all C atoms being equivalent; and the structure contains no pairs of adjacent pentagons which would provide an already activated point of attack for the addition of smaller carbon clusters. This contrasts with all other fullerenes in the cluster size range near C_{60}^{BF} (up to C_{68}) which must have at least one pair of adjacent pentagons. Thus in comparison with its isomers and neighbours C_{60}^{BF} once formed is more resistant to chemical attack than any other fullerene. C_{60}^{BF} is a survivor (Heath *et al.* 1987). Once formed, it should not react rapidly under most conditions. Thus any formation mechanism that has a significant fraction of the clusters reaching C_{60}^{BF} can account for the high yields because clustering can be effectively stopped at C_{60}^{BF}. No equilibrium between C_{60} and C_{70} need exist and the problem of relatively too much C_{70} encountered with the purely thermodynamic considerations above need not arise.

As we shall see, most proposed formation mechanisms aim at directly producing C_{60}^{BF}. Is this necessary? Below we show that thermodynamics strongly favours C_{60}^{BF} over other isomers of C_{60} and in the subsequent section we conclude that at elevated temperatures the ring rearrangements necessary to interconvert isomers can overcome kinetic activation barriers. This opens the way to formation mechanisms which merely have to avoid skipping over C_{60} too often.

Thermodynamic equilibrium favours C_{60}^{BF} over the other C_{60} fullerene isomers because it is much lower in energy. Presumably the next most stable isomer of C_{60} has two pairs of adjacent five-membered rings with C_{2v} symmetry. This has been calculated to be 193 kJ mol^{-1} (2 eV) higher in energy (Raghavachari & Rohlfing 1992). This C_{60} isomer is a closed, rigid cage with similar vibrational frequencies, and hence similar vibrational partition function. It is similar in overall size with similar rotational constants. Thus the internal partition function of C_{60}^{BF} is smaller than this next most stable isomer of C_{60} by roughly a factor of thirty from the ratio of symmetry numbers (2/60). The result is that the equilibrium constant for the reaction

$$C_{60}(\text{two fused pentagon pairs}) = C_{60}^{BF}, \quad (1)$$

is approximately 4000 at 2000 K. At lower temperatures the equilibrium will shift more towards C_{60}^{BF}. There are exactly 1812 fullerene isomers (Manolopoulos 1992; Liu *et al.* 1992), but the rest of these isomers will be higher in energy and should have negligible equilibrium populations.

6. Ring rearrangements on the fullerenes

Stone & Wales (1986) considered the ring rearrangement shown in figure 3*a*. They concluded that as a concerted process it has a Hückel four-centre transition state and thus will have a substantial activation barrier. The existence of such an activation barrier has been confirmed by the calculations of Yi & Bernholc (1992) who found activation energies in excess of 500 kJ mol^{-1} (5 eV).

However, a photofragmentation study of fullerene positive ions (O'Brien *et al.* 1988) provides some experimental evidence that ring rearrangements are reasonably facile. In these studies a positive ion was mass-selected, excited by an excimer laser, and the resulting fragment ions mass detected. Almost certainly these large ions

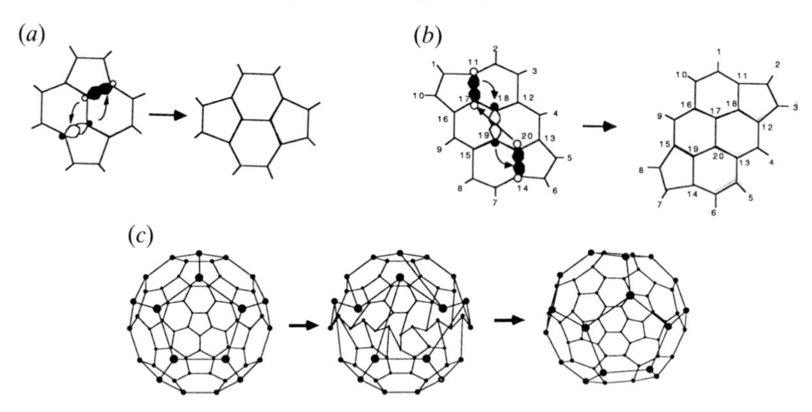

Figure 3. Schematic representation of some hypothetical fullerene ring rearrangement processes. (*a*) Stone–Wales rearrangement. (*b*) An allowed six-cycle rearrangement. (*c*) A global ring rearrangement on C_{60}. The reverse process is energetically favoured.

undergo rapid internal conversion so that the excitation energy is converted to vibrational energy. By varying the fluence of the excimer laser, the amount of energy deposited in the cluster can be varied. The time between excitation and observation of the charged fragment can also be varied. Figures 4 and 5 compare the fragmentation patterns with short and long times between excitation and observation. In figure 4, the time allowed for fragmentation is short (about 3 μs) and the fluence is high. In figure 5, the time allowed is longer (120 μs) and the fluence is lower. Thus the conditions of figure 4 favour unimolecular processes with large activation energies in contrast with figure 5 where unimolecular processes with lower activation energies are favoured. The loss of an even number of carbon atoms in the fragmentation is striking evidence that the cage structure is maintained. The fact that C_{60} is recognized as special in the observations at long fragmentation times is clear evidence that ring rearrangement processes must have a lower activation barrier than fragmentation.

There are other conceivable mechanisms for fullerene ring rearrangement in addition to the Stone–Wales process. Figure 3*b*, *c* shows some candidates. The Hückel six-centre transition state of figure 3*b* is an allowed concerted rearrangement, but the distortion required to reach the transition state is large. It should be noted that in figure 3*b* that atom 9 can be removed and atoms 15 and 16 linked converting their hexagon into a pentagon without really changing the mechanism. Likewise atom 4 could be removed. It is difficult to decide whether the cyclic rearrangement propagating around the molecule shown in figure 3*c* is an allowed concerted rearrangement. It is obviously a transition state in which many atoms move.

If ring rearrangements are facile at high temperatures, all that is required to obtain a high yield of C_{60}^{BF} is that the clustering should mostly pass through C_{60} and not skip over it, because the ring arrangements will lead downhill to C_{60}^{BF}. Indeed even if C_{60} is skipped over to C_{62} and perhaps even C_{63} and C_{64}, these clusters may dissociate back to C_{60}^{BF} at elevated temperatures. These dissociations remain endoergic (Stanton 1992), but because C_{62} has a much higher energy per C atom than C_{60}^{BF} they are less endoergic than typical fullerene dissociations, and entropy favours the dissociation. We will return the issue of whether C_{60} is likely to be skipped over in clustering.

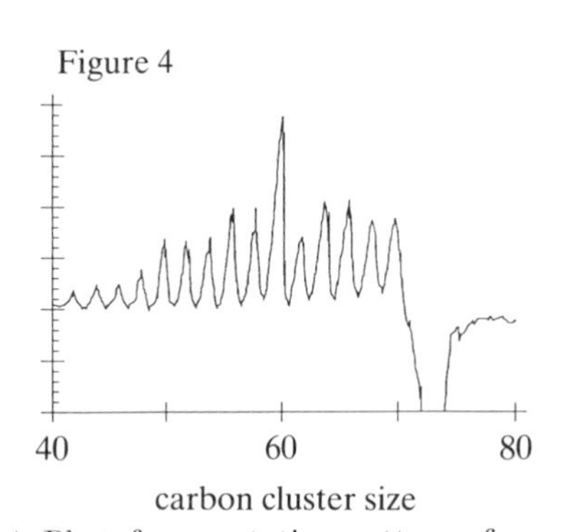

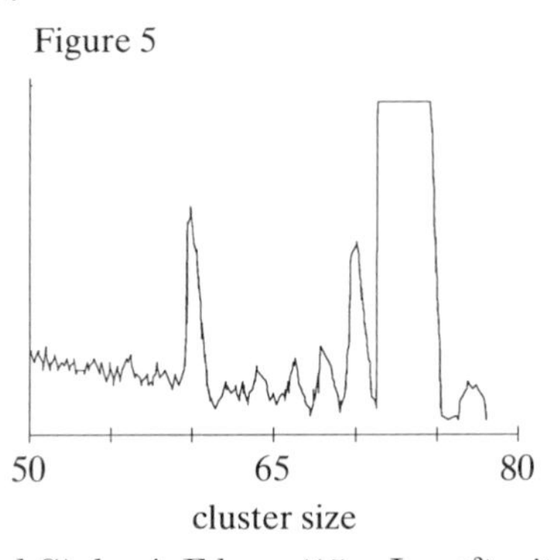

Figure 4. Photofragmentation pattern of mass-selected C_{74}^{+} by ArF laser (15 mJ cm^{-2}) with about 3 μs allowed between the ArF laser pulse and mass selection of the fragment ions. Note that C_{60}^{+} is only about twice as prominent as its neighbours. The C_{74}^{+} peak is downwards because the data presented is the difference between excimer laser on and excimer laser off.

Figure 5. Photofragmentation pattern of mass-selected C_{74}^{+} by ArF laser (15 mJ cm^{-2}) with about 120 μsec allowed between the ArF laser pulse and mass selection of the fragment ions. Note that C_{60}^{+} is much more prominent than its neighbours.

7. Carbon cluster kinetics and structures

The small target in phase space presented by the C_{60}^{BF} has led several investigators to propose special kinetic pathways leading directly to C_{60}^{BF}. Some general consideration of carbon clustering can provide a basis for examining the specific proposals.

A simple model of clustering kinetics is suggested by the phrase, 'everything that hits sticks'. One assumes that all clustering reactions are allowed, all are irreversible, and all have the same cross-section. If the clustering actually starts with atoms, this model is very unrealistic because the dimerization step, $2C \rightarrow C_2$, requires a three-body collision to stabilize the product. However, the saturated vapour over hot graphite consists largely of C_3 with C and C_2 the next most abundant species. The vibrational modes of the C_6 produced by dimerization of C_3 can temporarily store the energy released by formation of the new bond until the C_6 is stabilized by collision with the buffer gas; thus clustering can be initiated without three body collisions. Of the various refractory materials examined in cluster beam experiments, carbon clusters most rapidly because it starts with a primarily molecular vapour avoiding the three-body collision bottleneck (and because carbon forms very strong bonds).

Everything that hits sticks is readily integrated numerically. At intermediate clustering times, the resulting cluster distribution is approximately an exponential decay of cluster concentration with increasing cluster size. Major deviations from such a distribution can be interpreted as indications of size-dependent differences in reactivity.

Figure 6 depicts a carbon cluster distribution thought to be representative of a fairly early stage of clustering. It shows three distinct distributions consisting of some very small clusters with less than six atoms, an intermediate group from about seven to 22 atoms, and a heavy group from 36 atoms up. The first group, let us call it A, has both even and odd sizes and is dominated by the small clusters of three or less, the second group, B, has both even and odd cluster size and exhibits some rather marked almost periodic variations in intensity, and the last group, C, consists of only even clusters with C_{60} slightly more prominent.

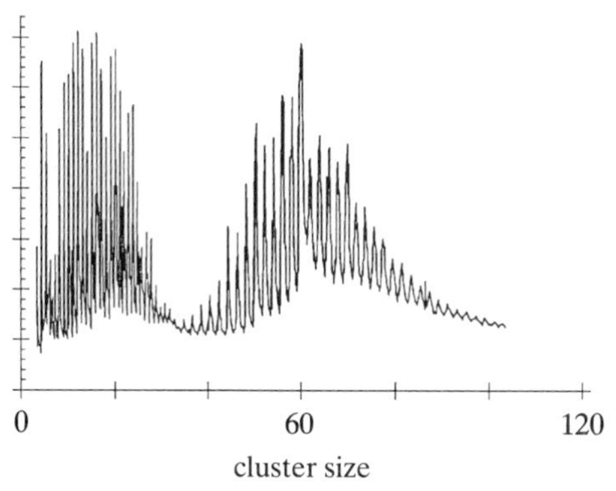

Figure 6. Typical carbon cluster distribution obtained in a laser vaporization supersonic beam source under relatively mild clustering conditions.

The most logical interpretation of the distribution is that clusters in the two low intensity regions, the 'gaps', are highly reactive while clusters in the high intensity regions are less reactive. The absence of odd clusters in the high mass region suggests that the odd clusters are much more reactive than the even ones. The high mass even clusters have been shown (Zhang *et al.* 1986) to be unreactive towards common reagents such as NO and are thought to be fullerenes. In laser cluster beam experiments with very little carrier gas, it is possible to produce cluster distributions near C_{60} with odd cluster intensities the same as those of the even cluster intensities with the C_{60} intensity hardly special at all. Presumably the carbon plasma expands so rapidly that clustering is stopped before the odd clusters react away. On the other hand, when clustering conditions are prolonged, high mass distributions, where C_{60} is essentially the only peak below C_{100} with appreciable intensity can be produced.

Recently, the ion drift tube mobility studies from Bowers' group (von Helden *et al.* 1991) have provided a means for separating carbon cluster ions with different structures because the reciprocal of the ion mobility is proportional to the collision cross-section. Thus a single cluster mass often consists of clusters with several different chemical structures and thus different cross-sections.

Such a mass-selected ion peak injected as a single packet into the drift cell separates into several peaks emerging from the cell at different times. The growth of collisional cross-section with size depends on the cluster shape, and thus in plots of inverse mobility versus cluster size the clusters can be grouped into chemical families which fall on a straight lines in the plot. Figure 7 shows such a plot of inverse mobility against cluster size for residual positive ion clusters from a supersonic excimer laser vaporization source. For a given cluster structure, a collisional cross-section can be fairly accurately calculated. Thus for a proposed structure a mobility can be calculated as a function of cluster size and compared with the mobilities of the families seen in figure 7. Thus a classification of components in the cluster beam by likely chemical structure emerges.

This classification scheme applies to positive *residual ions*, and we are interested in *neutral* clusters. Bowers' group (M. T. Bowers, personal communication) has also studied *residual anions* and find the essentially the same structural families but with quite different relative concentrations for a given mass and different ranges of cluster size where given families are important. Previous experience (O'Brien *et al.* 1986) indicates that the cluster distributions seen with residual *cations* is more similar to that of the *neutrals* than is the residual anion distribution. Thus the use of cation

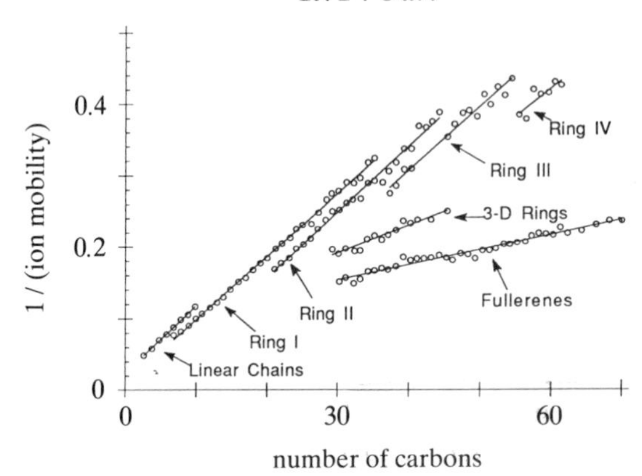

Figure 7. Reciprocals of the mobilities the carbon cluster cations as a function of cluster size from the work of Bowers and coworkers. The reciprocal of the mobility is proportional to collision cross-section which can be calculated from a model structure permitting the structural classification shown. Ring I is a single ring, ring II may be a planar bicyclic, ring III and ring IV may be one family and are guessed to be planar tricyclic rings. 3-D rings are guessed to be rings fused into a three-dimensional perhaps propeller-like structure.

distribution of figure 7 seems more appropriate to a discussion of the neutral distribution. The data in figure 7 were obtained under conditions where clustering was quenched at quite an early stage corresponding to the presence of odd peaks with intensities only somewhat less than the even peaks in the high mass region above 30 carbon atoms.

Comparison of figures 6 and 7, indicates that group A (the very small clusters of figure 6) must be linear, the second-group B must correspond to ring I, and the third group C must be, as already indicated, the fullerenes. Ring I species (B) and the fullerenes (C) have no dangling chemical bonds, thus accounting for their lack of reactivity. Ring II species have been assigned bicyclic structures and must be quite reactive as the ring II region overlaps the gap between 26 and 36 carbon atoms rather accurately.

For a given mass, von Helden *et al.* can measure the relative amounts of the various structures as they resolve into separate peaks in the drift cell. Under their clustering conditions, high mass clusters containing an odd number of carbon atoms are present in significant quantities implying that clustering has been quenched at an early stage. They find already C_{32}^{+} a few percent of the fullerene structure which grows to 98% by C_{60}^{+}.

8. Proposed kinetic models for C_{60}^{BF} formation

We have already seen that any model accounting for C_{60}^{BF} and $C_{70}(D_{5h})$ formation must be a kinetic one with major pathway(s) leading directly to C_{60}^{BF}.

(a) *Special pathways based on particular intermediate size clusters*

Those proposing a kinetic model in this sense believe that the potential energy hole of C_{60}^{BF} viewed as a target is too small to hit (at least often enough to account for the yield) by a chaotic clustering with many pathways. There are two models which combine several intermediate size clusters into the C_{60}^{BF} cage. A model of C_{60}^{BF} formation proposed by Goeres & Sedlmayr (1991) combines napthalenic C_{10} units.

Quantum theoretical calculations by Scuseria's group (Strout *et al.* 1992) show that napthalenic C_{10} is much higher in energy than simple C_{10} rings (ring I). This seems fatal for the model because no matter how reactive a species is, any pathway where it reacts with itself will be minor unless the concentration of the species is high. The other model of C_{60}^{BF} formation building up from intermediate size clusters has been proposed by Wakabayashi & Achiba (1992). In this mechanism napthalenic C_{10} combines with ring C_{18} to form a building block 'open' fullerene C_{28} well on the path to C_{60}. (The process continues C_{18}, C_{12}, C_2 to reach C_{60}.) As mentioned above, it is unlikely that napthalenic C_{10} can ever be present in significant amounts. However, the Scuseria group calculations (Strout *et al.* 1992) suggest that when ring C_{10} is placed inside a C_{18} ring, it begins to distort in the direction of producing the appropriate C_{28} intermediate.

An argument against mechanisms utilizing particular intermediate size clusters can be based on the fact that a wide variety of intermediate size clusters are observed in carbon cluster distributions which ultimately will produce larger clusters. Thus it is hard to account for the high yields of C_{60}^{BF} when these other intermediates pathways, which presumably lead to other products, are present.

(*b*) *The pentagon road*

To account for the formation of C_{60}^{BF} as a minor channel in supersonic cluster beams, we proposed the following picture (Heath *et al.* 1987). When the carbon cluster size passes about 30 atoms, graphitic sheets begin to be formed. As smaller species are ingested by the graphitic sheet, it attempts to follow the low energy path minimizing the number of dangling bonds by including pentagonal configurations at convenient places among the hexagonal ones. The incorporation of pentagons allows the sheet to minimize energy by curling up, reducing the number of dangling bonds. In this older picture, the pentagonal configurations introduced in this way will generally form rather randomly, giving rise to a multitude of curved shell-like graphitic networks of various shapes and sizes. Occasionally a sheet will 'accidentally' form in which the pentagons are so distributed so that closure can occur. The ion mobility work of von Helden *et al.* provides some support for this model in that ions are observed between 30 and 4 atoms with intermediate mobility between the fullerenes and the rings. These mobilities suggest a three-dimensional (3D) structure and the species were called 'open' fullerenes by von Helden *et al.*

To account for the observed high yields of C_{60}, this picture can be modified (Smalley 1992) by emphasizing the reduction in the number of dangling bonds through the incorporation of pentagons and de-emphasizing the random incorporation of pentagons. It is well-known (Schmalz *et al.* 1986) that adjacent pentagons are destabilizing. Construction of a carbon cluster, by specifying that it grows by addition of five-membered rings with closest vertices separated by a pair of six-membered rings on each side of the connecting bond between the five-membered rings, leads directly to C_{60}^{BF}. In this picture, open sheets are growing by addition of small clusters to their open edges with constant annealing so that the growth units described above are maintained. The driving force for following this pentagon road is that it reduces the number of dangling bonds in comparison with a hexagonal sheet while avoiding the high energy adjacent pentagon structures. In this model, growth takes place by following a minimum energy path in the broad valley of these pentagon road structures.

This model would be made more tangible if the 3D ring (or open fullerenes) seen in figure 7 could be identified with pentagon road structures, but obviously they cannot. By approximately thirty-six atoms the inverse mobility of the 3D ring species is already more than that of C_{60}^{BF}. The pentagon road species inverse mobilities would approach that of C_{60}^{BF} as their size approached 60. Thus there is no direct experimental evidence for the existence of pentagon road species in the structures found by ion mobility probing. However, it is important to realize the importance of a cluster structure to the main kinetic pathway is not measured by its concentration, but by the flux going through it. Thus a highly reactive species (and the pentagon road species are expected to be very reactive) may be present in very small concentrations and thus be very difficult to detect, but still carry most of the kinetics.

(c) *The fullerene road*

Heath (1991) has proposed that fullerenes form at cluster sizes of about forty and grow by the addition of small clusters, e.g. C_2 until C_{60} is reached. To have a growth mechanism which finds C_{60}^{BF} structure with its isolated pentagons, he proposed that the addition of small clusters to the fullerene cage proceeds in a manner which removes adjacent pentagons. This idea is attractive because adjacent pentagons are a high-energy site and thus are likely to be more susceptible to chemical attack. Heath proposed some specific mechanisms for attack by C_2 that remove adjacent pentagons. However, his mechanisms require rearrangements of the part of the fullerene cage not shown in his diagrams and it is not clear how reasonable these rearrangements can be made. Heath's picture of fullerene growth as the synthetic route to C_{60} can be modified by postulating that ring rearrangements take place more rapidly than growth and remove adjacent pentagons when possible as this lowers the energy.

A fullerene path picture of clustering leading to large C_{60}^{BF} yields emerges. The conditions needed when the cluster sizes approach the critical fullerene region are a moderately high temperature and a relatively low carbon density. As clusters sizes grow above 30 atoms, fullerenes begin to form. They grow in size either by additions of smaller clusters to the cage or perhaps by opening and reclosing. Odd clusters, of course, cannot have a fullerene structure, but are expected to be very fullerene-like in structure. When odd clusters are formed, they are expected to be very reactive and easily converted to fullerenes by loss or abstraction of a C atom. When a C_{60} cluster is formed, it is either a fullerene or almost a fullerene. Ring rearrangements on the C_{60} remove adjacent pentagons and lead downhill in energy until the C_{60}^{BF} isomer is found. By the arguments discussed above, once found C_{60}^{BF} is much less reactive and is effectively removed from the reaction scheme. The C_{60} atom cluster size is rarely skipped over because reactions where an intermediate size cluster adds to a fullerene require extensive reorganization of the fullerene, and therefore have a low rate. However, reactions in which an intermediate size cluster reacts with a fullerene to produce the next larger size fullerene expelling a smaller intermediate are not excluded. In C_{61}, C_{62} up to possibly C_{64}, there is some chance that the very hot cluster will lose C, C_2, C_3 or C_4 and revert to C_{60}; otherwise once past C_{60} the slide down the energy curve of figure 2 continues with another chance at C_{70} that the growing cluster can fall into the $C_{70}(D_{5h})$ potential well and be trapped.

Clustering at very high temperatures, as with high buffer gas pressures in graphite vaporization, is unfavourable to C_{60} formation because C_{60}^{BF} itself becomes more reactive under such conditions. If the temperature decreases too rapidly, as with low buffer gas pressures in graphite vaporization, clustering takes place so rapidly that C_{60} is often skipped over and the reorganizations, needed to find the C_{60}^{BF} minimum, do not compete with further clustering.

At the time that Heath proposed his fullerene mechanism, no macroscopic samples of endohedral metallofullerenes had been synthesized; he naturally argued that this fact supported the fullerene growth mechanism over the pentagon road. It is impossible to introduce a metal atom into a closed shell, whereas a metal atom could nestle into the cup formed by the pentagon road. However, since that time many metallofullerene species have been synthesized, and this fact appears, at least at first glance, to argue strongly against the closed fullerene growth mechanism. There are three ways that the fullerene road could be made compatible with the fact of endohedral metallofullerene synthesis: the metal atom attaches to the cluster before a closed fullerene is formed, the metal atom in some way can eat into a closed fullerene, or the synthesis of endohedral metallofullerenes is by a mechanism entirely different from that of the empty fullerenes. Of these possible escape routes for the fullerene growth mechanism, early attachment seems most plausible. In the pentagon path mechanism, as the cluster following the pentagon rule is closing, the doorway for entrance of the metal becomes smaller. Thus, it is unlikely that the metal attaches at the last stages if the pentagon road is the correct mechanism, and both mechanisms seem to require fairly early metal attachment.

9. Comparison of the two mechanisms

The pentagon road and the fullerene road are both fairly plausible mechanisms for the formation of C_{60}^{BF} in high yield in condensing carbon vapour. Both explain the need for an elevated temperature for effective fullerene formation: the pentagon road to bring about annealation to the isolated pentagon networks before further growth; the fullerene road to overcome activation barriers to fullerene growth and ring rearrangements. Both seem to require the presence of small carbon clusters for growth by addition of small clusters to growth elements in the 30+ to 58 carbon atom range, although each can be modified to eliminate the requirement for the presence of small clusters by adding mid-size clusters to the growing species with the rejection of part of the mid-size cluster.

The two mechanisms are qualitatively different. In the pentagon road mechanism, the chemistry is carried by highly reactive, high-energy species that are never present in very high concentration, but through which most of the chemistry passes much as combustion is carried by free radicals. Note that if growth of the pentagon road species involves very small carbon species, that both reagents have high energy and are highly reactive as in a radical–radical process. In the fullerene road, the fullerene reactants are the most stable clusters of their sizes and there is a great deal of experimental evidence that they are indeed present in high concentration. There is also evidence that they are reacting.

Under the wrong clustering conditions the C_{60}^{BF} yield can be very small. For example, buckminsterfullerene is not very prominent in the distribution shown in figure 6. Yet there is abundant evidence that the even number species larger than about 34 atoms are predominantly fullerenes in distributions where C_{60} is not prominent. These readily formed smaller fullereness disappear when clustering is prolonged at elevated temperatures leaving only C_{60} and C_{70} as survivors. There must be a reasonably facile mechanism by which closed fullereness can grow (and shrink) at high temperatures so that the smaller fullereness which are known to form can be destroyed and replaced by C_{60}^{BF} and larger fullereness.

It is known (Howard *et al.* 1991) that C_{60}^{BF} is produced in hydrocarbon flames in good yield. In the flame, the dangling bonds at the edge of a pentagon road sheets can be satisfied by H atoms removing the driving force to curvature. However, this observation is not fatal to the pentagon road mechanism, because the temperature is probably high enough to make the CH bonds labile and there are free radicals present capable of extracting H atoms from the sheet edge.

The chemistry of two mechanisms is very different, yet it is not easy to design an experiment that will reliably differentiate between them. Methods for synthesizing C_{60}^{BF} under more controlled conditions, such as the laser vaporization, heated flow tube of figure 1, and some means of *in situ* monitoring of the concentrations of the species present during growth are needed. Possibly the addition of small quantities of unsaturated hydrocarbons such as acetylene to the buffer gas flow would be illuminating since at 1200 °C HCCH would be expected to heal the dangling bonds on the edges of the growing sheets removing the driving force for curvature and stopping the growth. Acetylene might be less likely to hinder the growth of fullereness, but could hinder their formation.

This work was supported by the Robert A. Welch Foundation. I thank Richard Smalley, Thomas Schmalz and Michael Bowers for stimulating discussions which helped shape this work.

References

Goeres, A. & Sedlmayr, E. 1991 On the nucleation mechanism of effective fullerite condensation. *Chem. Phys. Lett.* **184**, 310–317.

Haufler, R. E. *et al.* 1991 Carbon arc generation of C_{60}. *Mater. Res. Soc. Proc.* **206**, 627–638.

Heath, J. R. 1991 Synthesis of C_{60} from small carbon clusters: a model based on experiment and theory. In *Fullerenes: synthesis, properties, and chemistry of large carbon clusters* (ACS Symp. Ser., no. 481) (ed. G. S. Hammond & V. J. Kuck), pp. 1–23. Washington, D.C.: American Chemical Society.

Heath, J. R., O'Brien, S. C., Curl, R. F., Kroto, H. W. & Smalley, R. E. 1987 Carbon condensation. *Comments Cond. Mater.Phys.* **13**, 119–141.

Howard, J. B., McKinnon, J. T., Makarovksy, Y., Lafleur, A. & Johnson, M. E. 1991 Fullerenes C_{60} and C_{70} in flames. *Nature Lond.* **352**, 139–141.

Krätschmer, W., Fostiropoulos, K. & Huffman, D. R. 1990 The infrared and ultraviolet absorption spectra of laboratory-produced carbon dust: evidence for the presence of the C_{60} molecule. *Chem. Phys. Lett.* **170**, 167.

Krätschmer, W., Lamb, L. D., Fostiropoulos, K. & Huffman, D. R. 1990 Solid C_{60}: a new form of carbon. *Nature Lond.* **347**, 354–358.

Kroto, H. W., Health, J. R., O'Brien, S. C., Curl, R. F. & Smalley, R. E. 1985 C_{60}: Buckminsterfullerene. *Nature Lond.* **318**, 162–163.

Liu, X., Schmalz, T. G. & Klein, D. J. 1992 Reply to comment on 'Favourable structures for higher fullereness'. *Chem. Phys. Lett.* **192**, 331.

Manolopoulos, D. E. 1992 Comment on 'Favourable structures for higher fullereness'. *Chem. Phys. Lett.* **192**, 330.

O'Brien, S. C., Heath, J. R., Curl, R. F. & Smalley, R. E. 1988 Photophysics of buckminsterfullerene and other carbon cluster ions. *J. chem. Phys.* **88**, 220–230.

O'Brien, S. C., Heath, J. R., Kroto, H. W., Curl, R. F. & Smalley, R. E. 1986 A reply to Magic numbers in C_n^+ and C_n^- distributions based on experimental observations. *Chem. Phys. Lett.* **132**, 99–102.

Parker, D. H., Wurz, P., Chatterjee, K., Lykke, K. R., Hunt, J. E., Pellin, M. J., Hemminger, J.

C., Gruen, D. M. & Stock, L. M. 1991 High yield synthesis, extraction and mass spectrometric characterization of fullerenes C_{60} to C_{266}. *J. Am. chem. Soc.* **113**, 7499–7503.

Pitzer, K. S. & Clementi, E. 1959 Large molecules in carbon vapor. *J. Am. chem. Soc.* **81**, 4477–4485.

Raghavachari, K. & Rohlfing, C. M. 1992 Imperfect fullerene structures: isomers of C_{60}. *J. phys. Chem.* **96**, 2463–2466.

Schmalz, T. G., Seitz, W. A., Klein, D. J. & Hite, G. E. 1986 C_{60} carbon cages. *Chem. Phys. Lett.* **130**, 203–207.

Slanina, Z., Rudzinski, J. M., Togasi, M. & Osawa, E. 1989 On relative stability reasoning for clusters of different dimensions: An illustration with the C_{60}–C_{70} system. *Thermochim. Acta* **140**, 87–95.

Smalley, R. E. 1992 Fullerene self assembly. *Acc. Chem. Res.* **25**, 98–105.

Stanton, R. E. 1992 Fullerene structures and reactions: MNDO calculations. *J. phys. Chem.* **96**, 111–118.

Stone, A. J. & Wales, D. J. 1986 Theoretical Studies of icosahedral C_{60} and some related species *Chem. Phys. Lett.* **128**, 501–503.

Strout, D. L. *et al.* 1993 (In the press.)

von Helden, G., Hsu, M-T., Kemper, P. R. & Bowers, M. T. 1991 Structures of carbon cluster ions from 3 to 60 atoms: Linears to rings to fullerenes. *J. chem. Phys.* **95**, 3835–3837.

Wakabayashi, T. & Achiba, Y. 1992 A model for the C_{60} and C_{70} growth mechanism. *Chem. Phys. Lett.* **190**, 465–468.

Yi, J-Y. & Bernholc, J. 1992 Isomerization of C_{60} fullereness. *J. chem. Phys.* **96**, 8634–8636.

Zhang, Q-L. *et al.* 1986 Reactivity of large carbon clusters: spheroidal carbon shells and their possible relevance to the formation and morphology of soot. *J. phys. Chem.* **90**, 525–528.

Discussion

R. C. Haddon (*AT & T Bell Laboratories, U.S.A.*). Presumably, as the fullerenes grow, by any of these mechanisms, they do so by one carbon at a time?

R. F. Curl. That is one of the questions to which we would like to know the answer. The small species you have present are C_1, C_2 and C_3 with, possibly, larger less reactive rings. I think growth occurs either by adding C_2, which is not too attractive, because it's a reactive species that is not present in high concentration, or by adding a C_3 atom and then 'spitting' out a carbon atom. In the fullerene growth mechanism you have to maintain the integrity of the cage when you add to it. You may add a C_3 molecule, form an imperfect cage, and then add another C_3 to form a perfect cage, if you use the species present in the highest concentration, C_3.

R. C. Haddon. So in that mechanism you actually avoid odd membered fullereness?

R. F. Curl. I would prefer to avoid them, but I am not sure they can be avoided, because the most logical reagent is C_3. You either have to spit out a carbon atom, which is a high-energy process, or you can tolerate the odd ones, which then turn out to be very reactive and immediately go on to the next step. The big problem that you have with the 'fullerene road' theory, is that you don't want to jump over C_{60}. Say you have large quantities of polyacetylene rings, and you take C_{56} and a 10 membered polyacetylene type ring, and you stick it on, you jump way over C_{60}. If this is to be a plausable mechanism, you have to have some way of sticking on part of a C_{10}, so that you don't jump over the big hole in the road that corresponds to buckminsterfullerene.

R. C. Haddon. Do you regard these intermediates as fluxional, so that the carbon atoms get totally redistributed throughout the molecule as they get added? Based on the isotope experiments, there has to be some sort of randomization going on.

R. F. Curl. I assumed in the isotope experiments that randomization was probably occurring in the vaporization step; that is, you obtain such small fragments on vaporization, that a random situation develops. It would be useful to think of an experiment where you started with a smaller fullerene, say all ^{12}C, and then brought in another which was all ^{13}C, and see how that randomized, but I think that the randomization in those processes may occur in the very first steps.

Production and discovery of fullerites: new forms of crystalline carbon

By Wolfgang Krätschmer[1] and Donald R. Huffman[2]
[1] *Max Planck Institut für Kernphysik, 6900 Heidelberg, P.O. Box 103980, Germany*
[2] *Department of Physics, University of Arizona, Tucson, Arizona 85721, U.S.A.*

Small carbon grains are assumed to be the carrier of the prominent interstellar ultra violet absorption at 217 nm. To investigate this hypothesis, we produced small carbon particles by evaporating graphite in an inert quenching gas atmosphere, collected the grains on substrates, and measured their optical spectra. In the course of this work – which in the decisive final phase was carried out with the help of K. Fostiropoulos and L. D. Lamb – we showed that the smoke samples contained substantial quantities of C_{60}. The fullerene C_{60} (with small admixtures of C_{70}) was successfully separated from the sooty particles and, for the first time, characterized as a solid. We suggested the name 'fullerite' for this new form of crystalline carbon.

1. Introduction

The production of laboratory analogues of interstellar grains was the initial aim of our research. In the autumn of 1982 while one of us (D.R.H.) was a Humboldt Fellow at the Max Planck Institute of Nuclear Physics in Heidelberg we decided to study the optical spectra of carbon grains. We felt challenged by the intense, strong interstellar ultra violet (UV) absorption at 217 nm which it had been proposed was due to graphitic grains (see, for example, Stecher 1969). The arguments in favour of such carriers are based primarily on calculations of the absorption of small, almost spherical, particles which exhibit the dielectric functions of graphite (for more recent literature see, for example, Draine 1988). There had already been very early experimental attempts to produce graphitic smoke particles by almost the same technique that we later applied to C_{60} production (see, for example, Day & Huffman 1973). The results provided some support for the graphite particle hypothesis, however, there were serious mismatches, especially with the width of the interstellar band (see Huffman 1977).

2. Carbon smoke particle production

First we repeated the older smoke production experiments, in which graphite rods were evaporated in a quenching atmosphere of helium. A conventional bell-jar carbon evaporator was used for this purpose. The carbon was vaporized by resistive heating, i.e. by passing a current through a pair of touching graphite rods. Helium pressures ranging between a few to *ca.* 20 torr were used to cool and to promote nucleation and aggregation of atoms and molecules in the carbon vapour into small solid particles. The smoke was collected at different positions within the bell jar,

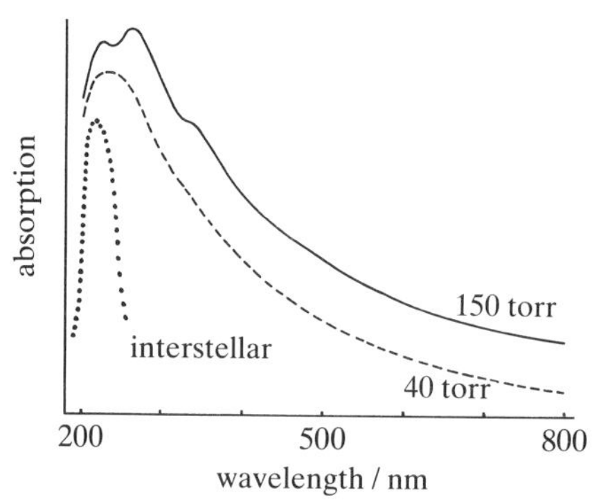

Figure 1. Absorption spectra of collected carbon particles produced by evaporating graphite in a quenching atmosphere of helium. The samples were obtained at different quenching gas pressures. Notice the occurrence of additional absorptions in the UV-visible at elevated helium pressures. These features turned out to originate from C_{60} (buckminsterfullerene). For comparison, the interstellar 217 nm absorption is also shown.

measured spectroscopically in the UV-visible range and examined by means of a transmission electron microscope. When the helium quenching gas pressure was chosen appropriately (*ca.* 20 torr), the spectra of the soot samples peaked at *ca.* 220 nm, i.e. at about the position of the interstellar feature. In all the spectra, however, the widths of the measured extinction curves were much greater than those calculated for graphite particles and also the observed interstellar feature (see figure 1). Part of this discrepancy may be due to the unavoidable clumping of the laboratory-produced particles, which to this day is a major experimental problem. Clumping increases the effective grain size and changes both the effective shape and optical properties of the particles. Another reason for the mismatch may lie in the carbon structure of the grains. Judging from Raman data, the particles we produced consisted of rather disordered graphite (see Huffman & Krätschmer 1990). Grains that have a more ordered graphitic structure are expected to show absorption features with considerably smaller widths.

It may appear that our smoke experiments did not yield much in the way of new insights, however, an interesting effect occurred which caught our attention. At our standard 20 torr helium quenching gas pressure we occasionally observed three additional features superposed to the continuum besides the 'regular' soot absorption feature (see figure 1). This three feature pattern appeared sometimes strong, sometimes weak, sometimes it was absent. Having no explanation, we whimsically called these additional features the 'camel humps'. Only a few observations were made on the carrier of the humps: the samples showed a peculiar Raman spectrum (Huffman & Krätschmer 1990), and the humps disappeared when the sample was heated in air. After much discussion we concluded that the carrier must be an artefact or 'junk'. This did not appear unlikely since the soot, with its high specific surface area, should be able to sweep up various contaminants, such as the pump oil or the vacuum grease which we used in the bell-jar evaporator. We then tried to produce unclumped grains, and in the following years turned our attention to matrix isolation studies of carbon. However, we kept the mysterious 'camel hump' features in mind.

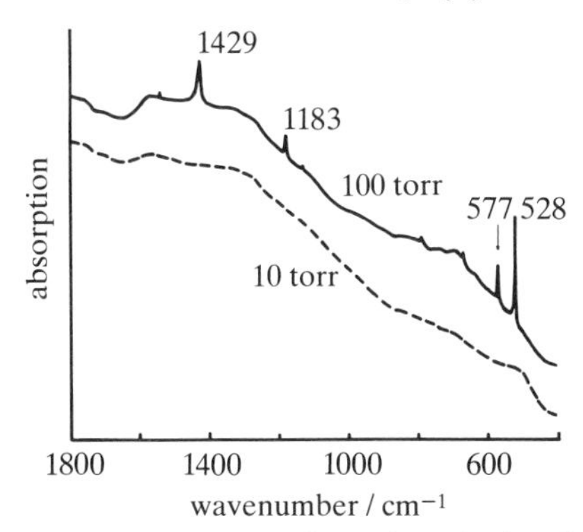

Figure 2. Absorption of carbon particles in the infrared. At elevated helium quenching gas pressures, there are four stronger lines emerging out of the continuum. Theory predicts precisely four ir-active fundamentals for a football structured C_{60} molecule and they should occur close to the measured positions. Such spectra gave us the first strong hint that we produced C_{60} in macroscopic quantities. Some of the tiny line features originate from the less abundant fullerent C_{70}.

3. Buckminsterfullerene

In the autumn of 1985 we read about the exciting C_{60}-Buckminsterfullerene discovery and about the ball proposal for its molecular structure (Kroto *et al.* 1985). Many theorists began to predict (among other things) its UV, infrared (IR), and Raman spectra. A remarkable result was that – for symmetry reasons – the infrared spectrum of football C_{60} should exhibit only four absorptions, two grouped at around 1600 cm^{-1}, and two grouped at around 600 cm^{-1} (see, for example, Weeks & Harter 1989). However, at this time the C_{60} molecules were only produced in cluster beams, i.e. in quantities insufficient for conventional optical spectroscopy or X-ray structural analysis. The football structure, even though 'too beautiful and perfect to be wrong' (Kroto 1992) remained a hypothesis.

In 1986 one of us (D. R. H.) began to consider that our camel hump carrier might possibly be due to the elusive buckminsterfullerene, and in 1987 a patent disclosure was submitted to appropriate channels of the University of Arizona. When we both met at an IAU Symposium in Santa Clara, California in summer of 1988, we discussed the case for C_{60} again. If the camel hump carrier is not 'junk' what else could it be other than C_{60}? Apart from a vague similarity between the camel hump features and the calculated uv spectrum of C_{60} (Larsson *et al.* 1987), however, the evidence was rather circumstantial. The few measured spectral data available on C_{60} did not support this idea (see Heath *et al.* 1986).

At the conclusion of the Symposium, we made new efforts to produce and investigate the camel hump soot. With the help of B. Wagner (in Heidelberg) and L. D. Lamb (in Tucson) we found the helium pressure to be the key parameter for producing camel hump features. Between 100 and 200 torr of helium the features appeared regularly, and we continued to use such pressures in all future work. In Heidelberg, the newly produced camel samples were studied with a very sensitive Fourier transform IR spectrometer. To our surprise, besides an intense continuum of regular graphitic smoke particles, four stronger line absorptions were clearly discernable, at frequencies very close to the positions predicted by vibrational theory for the football C_{60} (see figure 2). There was much excitement, since this implied that our samples not only contained C_{60}, but also that it was present in considerable quantities. We estimated the concentration as an order of a percent of the graphitic soot. However, the euphoria soon was followed by considerable concern. We still could not completely exclude the possibility that the IR lines were produced by contaminant, and we felt that such a sensational result, if true, had to be substantiated by additional experiments.

4. Fullerites

As part of his thesis work K. Fostiropoulos, in Heidelberg, investigated the effects of possible soot contamination and used much cleaner conditions and procedures than used previously. The appearance of the camel hump was not affected. With slightly renewed confidence we presented a paper of this observation at a workshop on interstellear matter in September 1989 (Krätschmer *et al.* 1990*a*). In a final attack on the problem (junk or not) Fostiropoulos and one of us (W.K.) succeeded in producing carbon rods that were sintered from 99% isotopically enriched ^{13}C powder. Shortly afterwards (late February 1990) the rods were successfully evaporated and the valuable smoke collected. We observed precisely the lineshifts in the IR spectrum of the soot as expected for pure carbon, while the UV spectrum remained unchanged. The conclusion was now clear: the carrier of the mysterious UV and IR lines had to be a pure carbon molecule and not junk. This and some additional results were written up and submitted for publication (Krätschmer *et al.* 1990*b*). From this time on, we coordinated our efforts and kept in constant contact by mail, fax and telephone. There was also more urgency to our work, since we noticed we were not alone in pursuing C_{60} and soot. The Sussex-group had succeeded in producing camel samples and confirmed that it exhibited four IR lines. There were also rumours that a group at IBM in San Jose was working in this direction.

At the end of April 1990 we were contacted by W. Schmidt, a chemist we knew from his work on interstellar dust and polycyclic aromatic hydrocarbon molecules. Following his suggestions, we achieved two important successes in May 1990 leading to a separation of the C_{60} from the soot. First, we discovered that heating the soot to *ca.* 500 °C *in vacuo* or in an inert atmosphere results in sublimed coatings which show all the absorption features of the camel samples, without any soot continuum. Second, we observed that the coatings dissolved in benzene, toluene, and various other non-polar solvents. This finding led naturally to the method of separating the insoluble graphitic grains from the fullerenes by simply washing the fullerenes out of the soot sample. A drop of the concentrated solution, dried on a microscope slide, provided the first view of the crystals of this new form of carbon, as shown in figure 3. One very easily obtains thin flake-like crystals which are ideally suited to electron diffraction studies. The crystal powder was subjected to X-ray Debye–Scherrer analysis. Mass spectrometry confirmed that the separated material consisted of C_{60} with a few percent C_{70}. From the diffraction data we deduced that the fullerene molecules form a lattice according to the close packing of spheres. The nearest neighbour distance turned out to be about 1 nm, which is close to the distance expected for the spherically shaped C_{60}, when a reasonable extension of the π-electron cloud is taken into account. These data, which strongly support the football structure of C_{60}, were submitted to *Nature* and published in September 1990 (Krätschmer *et al.* 1990*c*).

It was clear that for the first time we had produced a brand new crystalline form

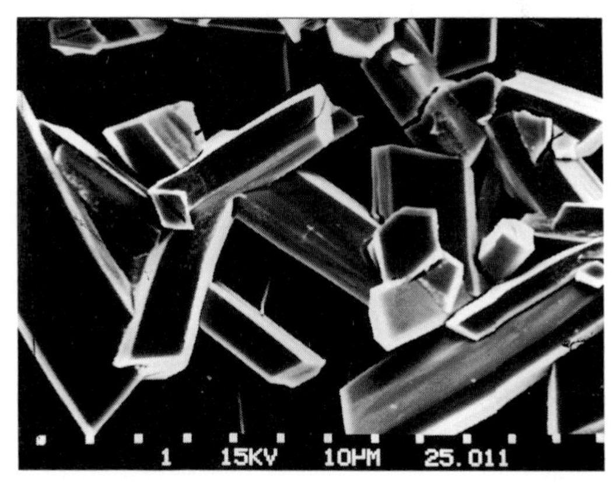

Figure 3. One of the first scanning electron microscope pictures of fullerene crystals, obtained by drying from a benzene solution. The picture was taken by K. Fostiropoulos. The rod-shaped crystals consist of C_{60} with a small admixture of C_{70} and traces of solvent molecules.

of pure carbon and that the molecular crystals are composed of the long-sought spherical molecule C_{60}, buckminsterfullerene. For this reason we suggested the special term *fullerite* for the solid.

In addition we had devised a method for making copious quantities of this molecule, discovered in 1985 but never previously made in sufficient quantity to be seen. In the work following our first studies, many other groups used the new material and soon it became proven beyond any doubt that the proposal by Kroto and co-workers of a ball shaped C_{60} molecule was correct. More recently, fullerene researchers started to study not only the soluble fractions of the soot, i.e. C_{60}, C_{70}, and the other larger stable 'magic number' fullerenes, but also they have looked very closely at the soot collected from various parts of the vacuum chambers and from the unevaporated residuals on the graphite electrodes. They found a variety of closed caged structures: round, elongated, tube-like, single-layered or nested within each other. It appears that a large scientific and technological playground with nicely shaped nano-toys has been opened. The challenge is to work out how to separate all these molecules, determine their properties and, finally, find uses for them.

References

Day, K. L. & Huffman, D. R. 1973 Measured extinction efficiency of graphite smoke in the region 1200–6000 Å. *Nature phys. Sci.* **243**, 50–51.

Draine, B. T. 1988 On the interpretation of the 2175 Å feature. In *IAU Symp. on Interstellar Dust* (ed. L. J. Allamandola & A. G. G. M. Tielens), vol. 135, pp. 313–327.

Heath, J. R., Curl, R. F. & Smalley, R. E. 1987 The UV absorption spectrum of C60 (buckminsterfullerene): A narrow band at 3860 Å. *J. chem. Phys.* **87**, 4236–4238.

Huffman, D. R. 1977 Interstellar grains. The interaction of light with a small-particle system. *Adv. Phys.* **26**, 129–230.

Huffman, D. R. & Krätschmer, W. 1991 Solid C_{60}: How we found it. In *Clusters and cluster-assembled materials* (ed. R. S. Averback, J. Bernholc & D. L. Nelson) (Symp. Proc. vol. 206). Pittsburgh: Material Research Society.

Kroto, H. W., Heath, J. R., O'Brien, S. C., Curl, R. F. & Smalley, R. E. 1985 C_{60}: Buckminsterfullerene. *Nature, Lond.* **318**, 162–163.

Kroto, H. W. 1992 C_{60}: Buckminsterfullerene, the celestial sphere that fell to earth. *Angew. Chem. Int. Ed. Engl.* **31**, 101.

Krätschmer, W., Fostiropoulos, K. & Huffman, D. R. 1990*a* Search for the UV and IR spectra of C_{60} in laboratory-produced carbon dust. In *Dusty objects in the universe* (ed. E. Bussoletti & A. A. Vittone), pp. 89–93. Kluwer Academic.

Krätschmer, W., Fostiropoulos, K. & Huffman, D. R. 1990*b* The infrared and ultraviolet absorption spectra of laboratory-produced carbon dust: evidence for the presence of the C_{60} molecule. *Chem. Phys. Lett.* **170**, 167–170.

Krätschmer, W., Lamb, L. D., Fostiropoulos, K. & Huffman, D. R. 1990*c* Solid C_{60}: a new form of carbon. *Nature, Lond.* **347**, 354–358.

Larsson, S., Volosov, A., Rosén, A. 1987 Optical spectrum of the icosahedral C_{60} fullerene-60. *Chem. Phys. Lett.* **137**, 501–503.

Stecher, T. P. 1969 Interstellar extinction in the ultraviolet. *Astrophys. J.* **157**, L125–L126.

Weeks, D. E. & Harter, W. G. 1989 Rotation-vibration spectra of icosahedral molecules. Icosahedral symmetry, vibrational eigenfrequencies, and normal modes of buckminsterfullerene. *J. chem. Phys.* **90**, 4744.

Discussion

M. Jura (*University of California, U.S.A.*). There has been a report from a shuttle experiment that the UV polarization of the interstellar extinction continued through the 220 nm feature. Is this consistent with the notion that since you would expect the polarization produced by grains, for example picket-fence molecules, to be lined up by a magnetic field, how can this occur if the molecules are spherical?

W. Krätschmer. That would not support our picture of course, but I wonder if this is really the case. There have been reports to the contrary, so we are waiting for the decisive result.

S. Leach (*Observatoire de Paris – Meudon, France*). Looking at hydrogen deficient stars, the peak in the UV spectrum shifts to about 250 nm. Do you think this means that hydrogen is involved in some way with the 230 nm peak?

W. Krätschmer. This may be the case. A shift can be produced in many ways. I have assumed the particles are small compared with the wavelength. If the particles are large, then you can have the peaks shifted to a longer wavelength. So one can assume that outflows from stars may consist of larger particles. On the other hand, one can not exclude the presence of hydrocarbons, for which there is plenty of evidence. But the spectra of these compounds are not well known as yet. These kinds of particles are all possible, and I would say that certain amounts of hydrogen will be involved, as hydrogen is everywhere. But, obviously the amount is restricted to such an extent that it has no major influence on the optical properties.

Systematics of fullerenes and related clusters

By P. W. Fowler

Department of Chemistry, University of Exeter, Stocker Road, Exeter EX4 4QD, U.K.

Qualitative theoretical treatments of the fullerene family of molecules can be used to count possible isomers and predict their geometric shapes, point groups, electronic structures, vibrational and NMR spectroscopic signatures. Isomers are generated by the ring-spiral algorithm due to D. E. Manolopoulos. Geometrically based magic number rules devised by the present author account for all electronically closed π shells within the Hückel approximation and these 'leapfrog' and 'cylinder' rules apply to the wider class of 'fulleroid' structures constructed with rings of other sizes. Extrapolations from the theory of carbon clusters are described for doped fullerenes, metallocarbohedrenes, fully substituted boron–nitrogen heterofullerenes and decorated-fullerene models for water clusters.

1. Introduction

The fullerenes offer a challenge to theoretical chemistry. They are large molecules and, even with modern computational methods, it would be expensive and often uninformative to perform full *ab initio* calculations on them cage by cage. In the first steps towards understanding these new materials a more qualitative approach is necessary and desirable. Methods based on little more than topology and elementary valence theory can provide information on isomerism, geometric and electronic structure, spectroscopic signatures, stability and selection rules for interconversion. This paper touches on developments in these areas but, in line with the 'post-buckminsterfullerene' theme of the meeting, concentrates on new ideas in the theory of exotic fullerenes, heterofullerenes and related heteroatom and molecular clusters.

2. Fullerenes: isomers and electronic structure

The discovery of C_{60} (Kroto *et al.* 1985) carried with it the implication of the existence of a whole series of similar molecules. In the original experiments strong C_{60} and C_{70} signals were found in association, and intensity was distributed over a wide range of even-numbered clusters. Theoretical rationalizations of the stability of C_{60} made in that paper and elsewhere postulated a pseudospherical σ and π system which clearly admitted application to other nuclearities (Fowler & Woolrich 1986; Fowler 1986; Schmalz *et al.* 1988). The term 'fullerene' was coined (Kroto 1987) and following the preparation of macroscopic quantities of C_{60} by a startlingly straightforward method (Krätschmer *et al.* 1990), higher members of the fullerene family are now being isolated in chemically tractable amounts. The current section concentrates on the relationship between isomer geometry and electronic structure.

A fullerene is a geometrically closed trivalent polyhedral network in which n carbon atoms are arranged in 12 pentagonal and $(\frac{1}{2}n-10)$ hexagonal rings. Such a

Phil. Trans. R. Soc. Lond. A (1993) **343**, 39–52

Printed in Great Britain

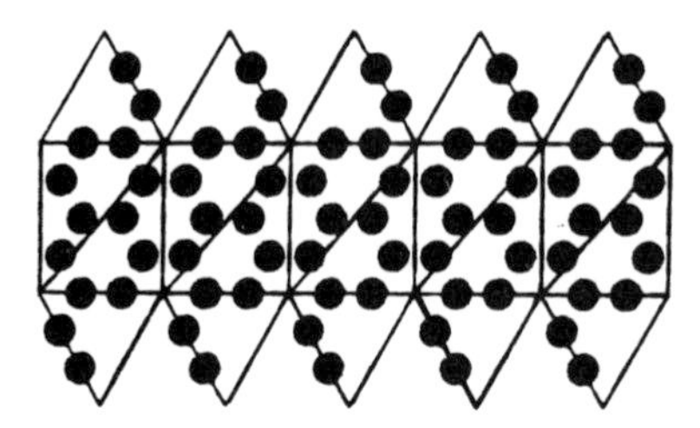

Figure 1. A net for C_{70} composed of five identical strips.

polyhedron can be constructed for all n greater than or equal to 20, with the single exception of 22 (Grünbaum 1967), and in the spherical approximation serves as the basis for a three-dimensional (3D) π system in which each carbon has three σ bonds to its neighbours and donates one electron to a π system of radial p orbitals. The number of isomers compatible with this definition grows rapidly with n.

An initial approach to fullerene enumeration was based on point-group symmetry (Fowler 1986; Fowler *et al.* 1988) and involved an extension of Coxeter's (1971) work on icosahedral tessellations of the sphere and of methods for the classification of virus structures (Caspar & Klug 1962). This approach led to magic numbers in fullerene electronic structure (Fowler & Steer 1987; Fowler 1990) and will be described briefly here.

The construction considers duals rather than the fullerenes themselves. Two polyhedra are dual to each other if the vertices of one coincide with the face centres of the other and vice versa. Each three-coordinate vertex of a fullerene corresponds to a triangular face in the dual, and the centre of each fullerene face to a five- or six-coordinate vertex of the deltahedral dual. The net of an arbitrary fullerene isomer is found by joining points in the infinite plane equilateral triangulated lattice. Twenty master triangles are drawn, joining 22 points of the plane by 19 internal and 22 external edges (figure 1). On folding the net, various vertices (and edges) collapse and a master icosahedron of 12 vertices and 30 edges is formed. The icosahedron is more or less irregular, with a symmetry that follows from the original choice of vertices and is the same as that of the dual fullerene. Its surface area is proportional to the number of small triangles cut from the plane and fixes the fullerene nuclearity. Hence, the enumeration problem reduces to finding all possible nets of given area compatible with a given point group, which in turn reduces to solution of integer equations.

For icosahedral symmetry the solution sets are well known from earlier work: An I or I_h fullerene C_n is possible for all n of the form $20[i^2+ij+j^2]$ (which implies either $n = 60k$ or $n = 60k+20$). The new observation was that electronic and geometric structures of the fullerenes are closely related: closed shells are found for the $60k$ series of which C_{60} is the prototype, but open (G^2) shells for the $60k+20$ series of which C_{20} is the first member (Fowler 1986). Nuclearities for tetrahedral fullerenes (T, T_d and T_h) are $n = 4[i^2+ij+j^2+k^2+kl+l^2+3(il-jk)]$ and for five- and six-fold cylindrical symmetry they are $n = 2p[k^2+kl+l^2+(il-jk)]$ with $p = 5$ and 6 respectively. Two- and three-fold dihedral symmetries were also treated. Details of the limitations on i, j, k, l and comprehensive lists of isomers are available (Fowler *et al.* 1988).

Systematic Hückel calculations on these and other series reveal the connection between geometry and electronic structure that is summarized in two rules. The more general, applying to all symmetries, is the leapfrog rule, which states that at least one isomer C_n with a properly closed π shell exists for all $n = 60+6k$ ($k = 0, 2, 3, \ldots$ but not 1) (Fowler & Steer 1987). In fact, there is one closed-shell C_n for every isomer of $C_{n/3}$, derived from it by the geometrical leapfrog construction. Starting from any fullerene structure, it is possible to construct one with three times as many atoms by first capping the faces and then taking the dual (figure 2). Whatever the starting isomer, the product fullerene always has a properly closed π shell (bonding HOMO, antibonding LUMO) in the simple Hückel approximation; the starting isomer has $20+2k$ ($k \neq 1$) atoms, hence the form of the rule. The fact that a closed shell is produced can be rationalized by considering local bonds and antibonds, spherical parentage of cluster orbitals or Kekulé structures (Fowler 1992), and can be proved by graph theory (Manolopoulos *et al.* 1992). The leapfrog rule subsumes the results for icosahedra, so that C_{60} is the first of an infinite leapfrog family.

The second, more limited, rule is for carbon cylinders. Within the five- and six-fold symmetric families described above, a single π closed shell (but with a non-bonding LUMO) is found for $n = 2p(7+3k)$ (all k). These isomers are expansions of C_{60} and their regular electronic structure follows from the solutions of the Schrödinger equation for the particle on a cylinder (Fowler 1990). C_{70} is the parent of this infinite family, and C_{84} is the first case for which both leapfrog (2) and cylinder (1) closed shells exist. The two rules cover all known cases of closed shells. Both constructions produce fullerenes in which the pentagonal rings are isolated from one another and so all properly closed-shell fullerenes obey the isolated pentagon rule (IPR), as do many open shell.

An independent attack on the enumeration problem was made by Manolopoulos *et al.* (1991) in Nottingham when they devised the spiral algorithm. This is based on the conjecture that any fullerene may be peeled as a coiled strip of pentagons and hexagons, so reducing the 3D structure of C_n to a 1D sequence of 12 pentagons interspersed amongst $(\frac{1}{2}n-10)$ hexagons. Fullerene graph enumeration is then a problem of generating all possible sequences of fixed length, discarding those that do not close to give a fullerene and eliminating the (many) redundant repetitions on the basis of their eigenvalue spectra. A spiral is easily 'inverted' to give the adjacency matrix. Approximate cartesian coordinates can be generated from eigenvectors of the adjacency matrix, (maximal) point group symmetry assigned, IR and NMR signatures of each isomer listed and an approximate electronic structure determined (Manolopoulos & Fowler 1992). Interconversions of isomers by motion of pentagons, and by C_2 intrusion and extrusion, have been mapped and used to explain statistical effects in experimental isomer distributions (Fowler 1992*a*, *b*; Manolopoulos *et al.* 1992*a*, *b*).

The spiral algorithm has been used to confirm that the 'leapfrog' and 'carbon cylinder' isomers are the only properly closed shells in Hückel theory. In turn the leapfrog construction provides informal support for the completeness of the spiral algorithm. In all cases examined, the listing of spirals for C_{3n} includes all the leapfrogs of C_n. The lists of isomers found from the tessellation for the higher point groups are in complete agreement with the symmetry analysis of the spiral lists. One motive for implementing this algorithm for all 28 possible fullerene groups is to give a further check on spiral completeness.

Given the efficient spiral algorithm for generation of isomer lists, many problems in structure and properties of higher fullerenes have been attacked. It is particularly important to have complete lists because it has turned out that possession of a closed

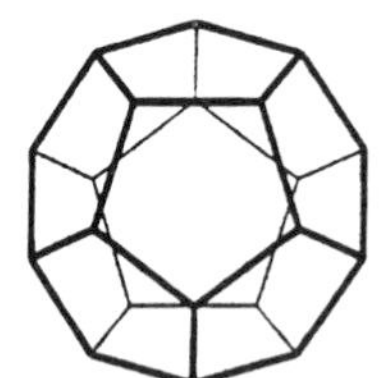
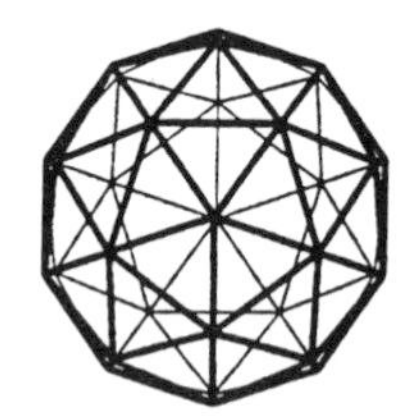
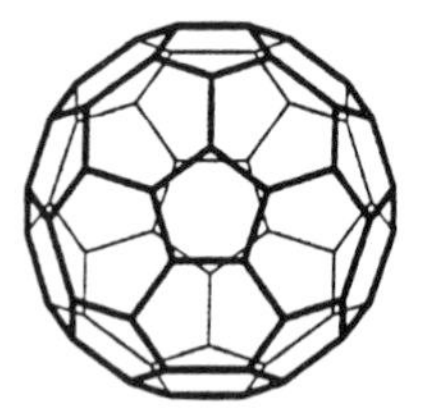

Figure 2. The leapfrog construction. A fullerene polyhedron is omnicapped and dualized to yield a fullerene with three times as many atoms.

shell is not the sole determinant of fullerene stability and electronic and steric factors are in fine balance. The experimental characterization of higher fullerene structures is providing some interesting surprises. All structures found so far are compatible with the IPR, and elimination of pentagon contacts has both steric and electronic advantages. However, local steric criteria such as the uniformity of the pattern of hexagon neighbours also seem important (Raghavachari 1992), and often favour structures of low point-group symmetry. Current beliefs on carbon cluster stability in this range of n can be summarized by a series of filters. First, discard non-fullerene isomers, then those with abutting pentagons, then isomers with disparate hexagon neighbour patterns and those with poor electronic structures. The balancing of the different criteria is not yet an exact science, but the emerging agreement between various qualitative and semi-empirical methods is encouraging. An extended discussion of the isomer problem, comparison with experiment and a catalogue of isomers, steric, electronic and symmetry data is to be published elsewhere (Fowler & Manolopoulos 1993).

3. Fullerenes and fulleroids

The fullerene structural pattern can be regarded as a minimal departure from the graphite template. Twelve pentagons introduce geometric closure, but otherwise all faces remain hexagonal. It seems at least feasible that other structures might have comparable stability, and in particular structures with just a few of their hexagons enlarged to heptagons might be worth theoretical investigation. Heptagons and octagons figure in the infinite lattices of the hypothetical graphite foams (Mackay & Terrones 1991; Lenosky *et al.* 1992), and heptagonal defects may account for the negative curvature observed in electron micrographs of carbon tubules (Iijima *et al.* 1992). Generation and motion of heptagonal defects have been studied by Saito *et al.* (1992). Large, negatively curved assemblies of carbon atoms have been proposed as building blocks for giant 'hyperfullerene' molecules (Scuseria 1992).

Staying close to the fullerene pattern, we have used qualitative molecular orbital theory to study clusters are similar to cylindrical fullerenes but with a p-gonal ring at each pole ($p = 7, 8, 9, \ldots$) and an extra $(2p-12)$ pentagonal faces (Fowler & Morvan 1992). The tessellation construction described earlier also applies to these 'fulleroid' cylinders, and so there is a seven-fold, eight-fold, ..., axially symmetric cage for every five- (and six-) fold symmetric fullerene. One strip repeated p times gives a net that wraps to form the three-dimensional pseudocylinder. For example, the same strip is used five times to give icosahedral C_{60}, six times to give D_{6d} C_{72}, seven times to give D_{7d} C_{84}, and so on. Corresponding to each five- or six-fold cylindrical isolated-pentagon fullerene is an isolated-p-gon fulleroid in which all defects are surrounded by hexagons.

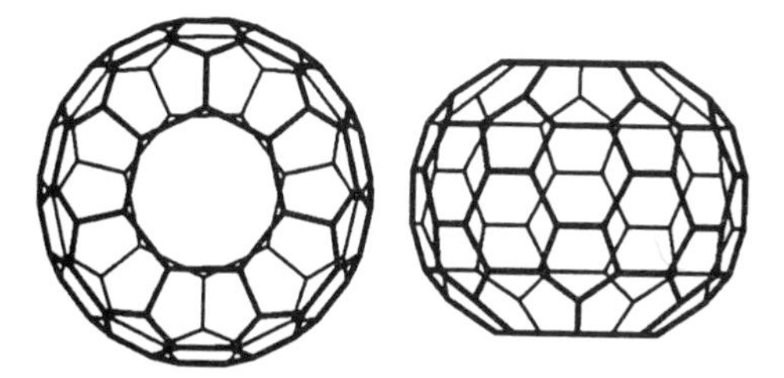

Figure 3. Two views of a hypothetical D_{7d} fulleroid isomer of C_{84}.

The electronic structure of the hypothetical fulleroids also parallels that of the fullerenes in several significant respects. Properly closed π shells are found within the Hückel approximation only for isolated p-gon isomers, and then only when the structure is either (*a*) leapfrogged from a smaller fulleroid of the same symmetry, or (*b*) a p-fold analogue of one of the fullerene 'carbon cylinder' series (Fowler 1990). D_{7d} C_{84} is the smallest fulleroid with a properly closed shell (figure 3).

Explicit calculation shows that leapfrogging of cages in this particular series produces a closed shell, but in fact leapfrogging of any trivalent cage with at least one pentagonal face will do the same (Manolopoulos *et al.* 1992). If the fulleroid class is defined as comprising those trivalent polyhedra with five or more atoms per face, then all leapfrog fulleroids have closed shells and the $60+6k$ magic number rule is extended accordingly.

The carbon-cylinder analogues obey a slightly more complicated rule. For fullerenes one closed shell below a non-bonding LUMO was found for nuclearities $n = 70+30k$ (five-fold symmetry) and $84+36k$ (six-fold symmetry). This generalizes to $n = 2p(7+3k)$ for the fulleroid cylinders, but analysis of the behaviour of the frontier orbital energies with ring size shows that, whereas k can take values $0, 1, 2, \ldots$ for $p = 5$ and 6, its range is restricted to the strictly positive integers for the higher values of p.

A measure of the electronic stability of the fulleroids is the π delocalization energy per atom. This energy is an increasing function of n for closed-shell cages, flattening off with large n (see fig. 3 in Fowler & Morvan 1992). As might be expected, the loss of hexagons leads to lower stability than for a closed-shell fullerene with an equal number of carbon atoms, but as the cluster size increases the π energy per atom tends to the graphite limit. The steric factor in overall stability is less easy to quantify, but unless the cluster is very large indeed there is likely to be considerable local strain around the heptagonal defects. MNDO calculations on D_{7d} C_{84} by Raghavachari & Rohlfing (1991) found it to be highly strained in spite of a large band gap. Prospects for fulleroid microfibres are brighter than for single-shell clusters.

4. Doped fullerenes and the metallocarbohedrenes

There is mass spectroscopic evidence for clusters $C_{59}B$, $C_{58}B_2, \ldots C_{54}B_6$ in the products of laser ablation of boron nitride/graphite composites (Guo *et al.* 1991). It is proposed that these may be versions of the fullerene cage in which heteroatoms have occupied carbon sites with little disturbance of the framework. This is a plausible hypothesis for low doping ratios, in view of the ability of carbon to substitute for boron in carboranes. Bond-energy considerations suggest that

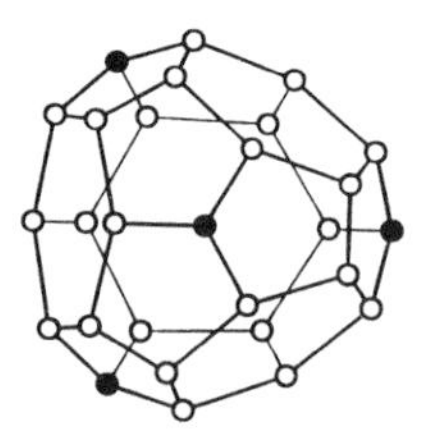

Figure 4. A hypothetical T_d isomer of $C_{24}N_4$ based on the C_{28} fullerene.

heteroatom sites would be well separated to maximize BC and minimise BB contacts, and the fact that each boron site takes up one ammonia molecule in titration experiments bears this out. Moreover, depending on the notional starting fullerene isomer, it is possible to imagine reinforcing or competing steric and electronic effects on isomer stability. Boron atoms would be likely to occupy planar sites (junctions of three hexagons) with low charge in the isoelectronic fullerene cation, whereas nitrogen substituents would be stabilized in pyramidal sites (junctions of three pentagons) with high negative charge in the isoelectronic fulleride anion. For example, substitution of N at the corner sites of C_{28} (figure 4) would give a closed-shell molecule isoelectronic with $C_{28}H_4$ and having four pyramidal nitrogen atoms.

More intricate theoretical issues emerge when we move one step further from the fullerenes to the newly proposed metallocarbohedrenes. Again at the time of writing the evidence is purely mass spectroscopic, but with the C_{60} precedent in mind we can hope for eventual confirmation by other techniques. It appears from laser ablation studies made by Castleman and co-workers using metal surfaces in contact with hydrocarbons that M_8C_{12} (M = Ti, Zr, Hf, V) has special stability, and magic number series M_nC_m have been interpreted in terms of growth of networks of fused dodecahedra (Guo *et al.* 1992*a*, *b*; Wei *et al.* 1992*a*, *b*). These clusters are exciting from the theoretical point of view as a possible embodiment of a substitutional analogue of the Jahn–Teller effect (Ceulemans & Fowler 1992).

The working hypothesis used in the experimental studies for rationalizing the stability of Ti_8C_{12} is that it is a geometrically and electronically closed dodecahedral framework notionally derived from the C_{20} fullerene by isolobal C → Ti substitution. Icosahedral C_{20} itself has an open-shell π configuration with two electrons in a fourfold degenerate non-bonding set of orbitals (Fowler & Woolrich 1986), and whereas the dication would have a closed shell, the neutral would be expected to undergo distortion to a geometry of lower symmetry in which double occupation of the lowest component of the symmetry-split frontier orbital set would yield a closed-shell singlet. The epikernel principle (Ceulemans *et al.* 1984) guarantees that the loss of symmetry will be the least possible that achieves the splitting, and for an I_h molecule with an open G shell the two maximal epikernel groups are T_h and D_{3d}.

The substitutional analogue of the Jahn–Teller effect on geometry arises as follows. Instead of geometric distortion lifting the degeneracy of the frontier orbitals, a splitting is induced by changing the chemical nature (and hence energetic parameters) of some of the atoms. This will drive further geometric changes too, as a simple effect of size (just as in the usual case the geometric changes induce changes in bond energetics). The epikernel principle governs the symmetries of effective substitution patterns; and both T_h and D_{3d} support substitution by eight atoms. In T_h the metal atoms occupy the corners of a cube; in D_{3d} they form pyramids capping a puckered C_{12} ring (figure 5).

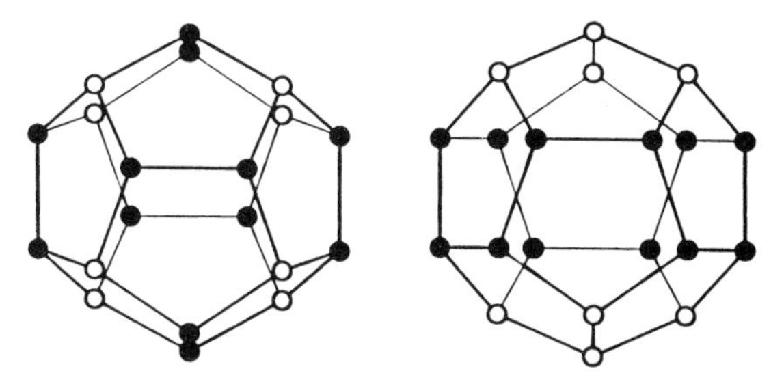

Figure 5. Tetrahedral and trigonal isomers of Ti_8C_{12}.

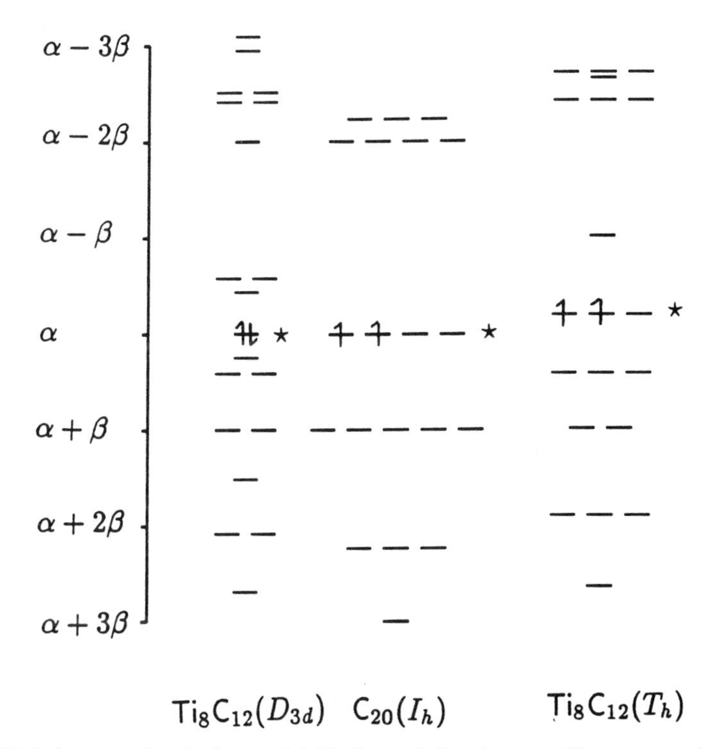

Figure 6. Hückel energy levels for model Ti_8C_{12} and C_{20} clusters. The star marks the HOMO and arrows indicate its occupation.

The induced splitting pattern can be deduced by first-order perturbation theory (Ceulemans & Fowler 1992). In the simple Hückel approximation, the electropositive metal atoms use sd hybrids for π bonding and have smaller negative α parameters. The relation $\alpha_{Ti} \approx \alpha_C - \beta_{CC}$ is suggested by differences in electronegativity, and changes in the β bond parameter have no effect in first order on the energy of the (non-bonding) G_u orbitals. Perturbation theory or explicit diagonalization show that in T_h symmetry the frontier level splits $G_u(4) \rightarrow A_u(1) + T_u(3)$ with the nondegenerate component higher in energy, and in D_{3d} the splitting pattern is $G_u(4) \rightarrow A_{1u}(1) + A_{2u}(1) + E_u(2)$ with the non-degenerate A_{1u} level at the bottom of the stack, at exactly the carbon $2p$ energy (figure 6). Therefore, if Ti_8C_{12} is a closed-shell singlet, it cannot have the T_h symmetry assumed by Castleman and co-workers, and the most symmetrical structure compatible with a closed shell is the D_{3d} cage proposed in our paper. As discussed there, the trigonal cage has a favourable aromatic carbon ring around its equator and 12 TiC contacts. The T_h structure has only isolated CC units but 24 TiC bonds. The two epikernel cages represent the extremes of localized (T_h) and delocalized (D_{3d}) electronic structures, and the experimental geometric structure will be a result of competition between the σ and π energies.

The main point illustrated by this over-simplified discussion of the prototype metallocarbohedrene is that a structure and electron count which is unfavourable for carbon may become electronically favoured by substitution of heteroatoms. The 20-electron π system of neutral C_{20} is the smallest of many potential examples of this effect in the fullerene series. An infinite number of icosahedral fullerene isomers with $60k+20$ atoms have the G^2 open-shell configuration as neutrals. Icosahedral C_{80} has a bonding HOMO with energy $\alpha+0.274\beta$ in the simple Hückel approximation. In I_h symmetry its atoms fall into sets of 20 (type 'h', common to three hexagons) and 60 (type 'p', common to two hexagons and a pentagon). In T_h symmetry, these split further into 8(h)+12(h)+12(p)+24(p′)+24(p″) (where p′ atoms are bonded to an atom in 8h, and p″ to an atom in 12(h)). Separate substitution of these sets by electropositive centres and calculation of the π energy levels shows that T_h isomers of $Ti_{12}C_{68}$ (with 12(h) = Ti) and $Ti_{24}C_{56}$ (with 24(p′) or 24(p″) = Ti) would have closed shells with small HOMO–LUMO gaps. For Ti_8C_{72} a D_{3d} isomer would have a closed shell, as in the 20-atom example.

5. Heterofullerenes and cages with isolated pairs of pentagons

Many chemists, familiar since schooldays with the crystal structures of graphite and hexagonal boron nitride must have asked themselves on seeing the C_{60} truncated icosahedron whether a $B_{30}N_{30}$ analogue would be possible (see, for example, Thompson 1992), and quickly rejected the idea because C_{60} is not an alternant. The stability of boron nitride arises from the strength of the heteronuclear BN bond, but it is impossible to alternate B and N around a pentagonal ring. Though pure boron clusters have been produced in laser ablation experiments (La Placa *et al.* 1992), there is no compelling evidence that they adopt fullerene structures, and for B_n a deltahedral shape seems more likely *a priori*. For a skeleton of a $(BN)_{n/2}$ molecule the duals of triangular tessellations of the octahedron (Goldberg 1937) may offer useful prototypes; these polyhedra have solely square and hexagonal faces. However, some variations on the fullerene theme also merit consideration.

The problem with BN occupation of a fullerene framework is not only the homonuclear bond forced in each pentagon, but the fact that the defect spreads into neighbouring rings (figure 7). An isolated-pentagon (BN) heterofullerene has at least 12 homonuclear contacts, but may have many more. If, however, the pentagons are allowed to pair up, the defect is contained, and the structure need only have six homonuclear contacts, one at the waist of each pair. Further aggregation of pentagons introduces more problem bonds, and so the isolated pentagon-pair (IPP) structure is optimal in that it minimizes the number of BB and NN links in a fully substituted fullerene.

Hypothetical B_xN_y cages based on IPP polyhedra have 6 BB, 3 BB+3 NN, or 6 NN homonuclear bonds, and overall stoichiometries $B_{n/2+2}N_{n/2-2}$, $B_{n/2}N_{n/2}$, $B_{n/2-2}N_{n/2+2}$, with $(n-4)$, n or $(n+4)\,\pi$ electrons respectively. This is proved as follows. Associate one third of each atom with each of its three bonds. In the total atom count, homonuclear BB bonds then contribute $\frac{2}{3}B+0N$ each, heteronuclear bonds contribute $\frac{1}{3}B+\frac{1}{3}N$, etc. Therefore the number of homonuclear bonds of BB type is

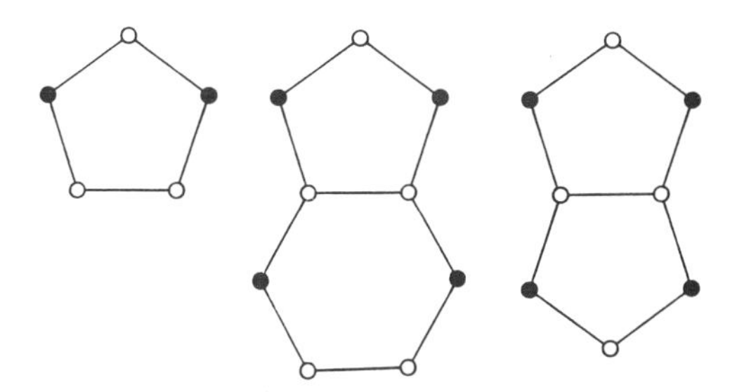

Figure 7. A pentagon has one forced homonuclear contact, which forces a second in a neighbouring hexagon. A pentagon pair has only one forced homonuclear bond which is without effect on neighbouring rings.

0, 3, 6. IPP fullerene polyhedra are labelled even or odd according to the parity of the number of BB links.

An NN contact would presumably provoke elimination of N_2 from the cage, and therefore a secondary stability condition is that all forced homonuclear links should be BB rather than NN. This can only be achieved by having an excess of four B atoms. A $(BN)_{n/2}$ IPP isomer must have three unfavourable NN contacts.

The two constraints define a class of polyhedral structures for which surface π and weakened edge σ bonding are possible. If the BB link is very weak it may be energetically favourable to break it altogether, leading to a true alternant polyhedron. Each pentagon pair would then become a distorted octagon and the structure would have $(\frac{3}{2}n-6)$ heteronuclear contacts, $(\frac{1}{2}n-10)$ hexagonal faces and six octagonal rings. It would also have 12 reactive two-coordinate atoms. In the original structure B forms three edge bonds with sp^2 hybrids and contributes an empty 2p orbital to the π system. With sufficient distortion at the low-coordinate sites, sp^2 hybridization could be retained, but with one hybrid now taking part in the π system and an empty B 2p orbital aligned with the deleted pentagon–pentagon edge. This latter bonding scheme would place an extra 12 electrons in the π system, with eight in non-bonding or nominally antibonding orbitals.

Irrespective of the viability of the bonding schemes outlined above, the mathematical problem of cataloguing the fullerene isomers that obey the starting condition of isolation of pentagon pairs is straightforwardly solved. One method is based on the spectral moments

$$\mu_j = \sum_i^n (\epsilon_i)^j$$

of the eigenvalues of an adjacency matrix (see, for example, Burdett *et al.* 1985). Through the association of each μ_j with closed walks of length j, the moments can be used to count structural components of a fullerene graph. A general fullerene C_n has moments $\mu_0 = n$, $\mu_1 = 0$, $\mu_2 = 3n$, $\mu_3 = 0$, $\mu_4 = 15n$, $\mu_5 = 120$, $\mu_6 = 93n-120$, and $\mu_7 = 1680$ but terms beyond μ_7 depend on the detailed arrangement of pentagons. A pentagon pair introduces an eight-ring and so modifies μ_8. Structures in which all pentagons are either isolated or linked in p pairs can be accommodated by the equations $\mu_8 = 639n-1920+16p$, $\mu_9 = 18360-36p$, $\mu_{10} = 4653n-22680+400p$, $\mu_{11} = 184800-924p$. Any further linking of pentagons into triples, etc., changes μ_9, μ_{10} or μ_{11} and can be excluded on the basis of these moments. Coupling the moment formulas to the Nottingham spiral algorithm (Manolopoulos *et al.* 1990) produces the IPP isomer counts listed in table 1. Six pentagon pairs account for 48 atoms, and it

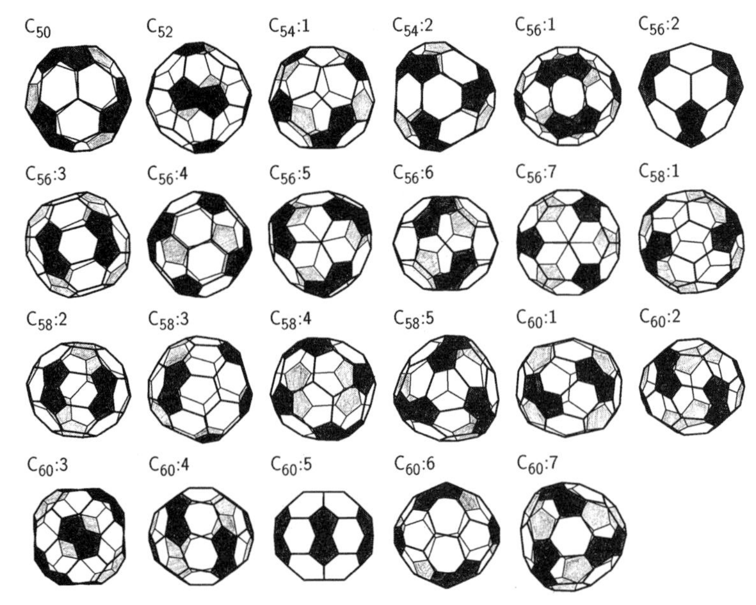

Figure 8. The isomers of C_n ($n \leqslant 60$) with six isolated pairs of pentagons. Symmetries are listed in table 1.

Table 1. *Fullerene isomers* C_n *with six isolated pairs of pentagons*
(Symmetries are listed for isomers in the order in which they are generated by the spiral algorithm. Even cages are marked by a star.)

n	count	point group symmetries
50	1	D_3
52	1	$T*$
54	2	C_2, D_3
56	7	$D_2, C_{2v}, C_2, C_2, D_3, D_2, D_{3d}$
58	5	C_2, C_2, C_1, C_2, C_3*
60	7	$C_2, C_2, D_{2d}, D_2, D_{6h}, D_2, D_3$
62	7	$C_2, C_2, D_{3h}, C_2, D_3, D_3, D_3$
64	11	$C_2, C_1, D_2, C_2, C_3*, C_2, C_2, C_1, C_2, C_2, D_2$
66	11	$C_2, C_1, C_2, C_1, C_1, C_2, C_2, C_2, C_2, C_s, D_3$
68	25	$C_2, C_2, D_3, C_2, C_1, C_1, C_2, D_3, D_2, C_2, C_1*, C_2, C_2, C_1, C_2, C_2, C_2,$ $C_2, C_2*, S_6*, C_2, D_2, C_3*, D_2, T_d*$
70	19	$C_2, C_1, C_2, C_1, C_1, C_1, C_{2V}, C_2, C_2, C_2, C_2, C_2, C_3*, C_1, C_2,$ C_2, C_2, C_2, C_2
72	34	$C_2, C_2, C_1, C_2, C_2, C_1, C_1, D_2, C_2, C_2, C_2, C_2, C_2, C_2, C_1, C_2, C_1,$ $C_1, C_2, C_2, C_1, D_3, C_2, C_2, D_6, C_2, C_2, C_1*, C_2, C_2*$ C_s*, D_2, C_s*, D_3

turns out that the smallest realizable IPP isomer has just two atoms outside pentagons and is a D_3 C_{50}. As with isolated-pentagon fullerenes, the number of IPP isomers grows erratically at first and is non-zero for every even vertex number above a threshold value. The first 23 IPP clusters are illustrated in figure 8.

The second criterion for stability implies that even polyhedra should be preferred to odd ones. One way to check the parity label automatically is to calculate the eigenvectors of the adjacency matrix obtained by deletion of the six pentagon–pentagon links of the original fullerene. Since the cut fullerene is an alternant, it has paired eigenvalues; the lowest non-degenerate vector is fully bonding and has all n coefficients in phase, and its totally antibonding counterpart has a phase change along each remaining edge. If the coefficients in the latter vector are of one sign for all 12 special atoms, then the cage is even.

The D_3 C_{50} cage is odd, as are most clusters examined, but the C_{52} is even and a total of 13 even structures are found amongst the first 130 IPP fullerenes (table 1). Simple Hückel calculations on the 52 vertex $B_{28}N_{24}$ cage with $\alpha_B \approx \alpha_C - \beta_{CC}$ and $\alpha_N \approx \alpha_C + x\beta_{CC}$ ($x = 1$ or $\frac{3}{2}$) show a closed shell at 48 π electrons with a large band gap, indicating stability for the neutral molecule with intact BB bonds. Semi-empirical calculations to determine the electronic structure and steric stabilities of more of these novel clusters are needed, and should give an indication of whether the polyhedra of figure 8 are chemically realistic or merely mathematical curiosities.

6. Decorated fullerenes and hydrogen bonding

The extrapolations from fullerenes described so far are models for chemically bonded frameworks of atoms, but the structural motif of hexagonal and pentagonal rings of linked units has much wider application in chemistry and beyond (Caspar & Klug 1962; Coxeter 1971; Marlin 1984). Given the prevalence of rings of corner-linked tetrahedra in silicates, it seemed logical to propose stability for fullerene-like silicate frameworks such as $Si_n O_{5n/2} H_n$ where each vertex of the fullerene polyhedron is 'decorated' by a tetrahedron composed of a central Si atom joined to three others by Si–O–Si linkages, and carrying an *exo* OH group (Fowler *et al.* 1991).

Another type of decorated fullerene discussed in the same paper has already been detected in experiment. This is the hydrogen-bonded fullerene-like water cluster. Drawing on the classical picture of the water molecule as a tetrahedral oxygen capable of participating in four hydrogen bonds, two as proton-donor and two as acceptor, a finite 2D ice-like network can be constructed on the surface of a hollow sphere. In this decoration each molecule is linked to three neighbours by hydrogen bonds, and the fourth vertex of the tetrahedron is occupied by either H or a lone pair. It can be shown that at least one closed structure $(H_2O)_n$ with $\frac{1}{2}n$ *exo* H atoms and $\frac{1}{2}n$ *exo* lone pairs can be constructed for every isomer of the n-vertex fullerene. The simplest of these is based on the Platonic dodecahedron (i.e. C_{20}). With an embedded (and possibly mobile) proton or hydroxonium ion, this cage had already been proposed as the explanation of $(H_2O)_{20}H^+$ and $(H_2O)_{21}H^+$ clusters observed in mass spectra of water vapour (Kassner & Hagen 1976; Wei *et al.* 1991). Structures equivalent to decorations of the 24-, 26- and tetrahedral 28-vertex fullerenes enclose Cs^+ and $H_2O \ldots Cs^+$ in proposals for gas-phase clathrates made by Selinger & Castleman (1991). Simulation of water clusters by Monte Carlo and other methods produces $(H_2O)_{20}$ and $(H_2O)_{60}$ cage structures topologically equivalent to decorated fullerenes, though with low point group symmetry. For a given fullerene frame, many conformers of similar energy may exist, and internal motion or proton conduction mechanisms are easily envisaged. A clear difference between water clusters and carbon fullerenes is that the interaction between units is σ in character, directed along the edges of the polyhedron. Lacking the π interactions of carbon rings, the faces are not constrained to near planarity and the geometries may depart considerably from the idealised model. The decorated fullerene series does however define a systematic way of generating starting configurations for water clusters, solvation cages and microstructures in liquid water.

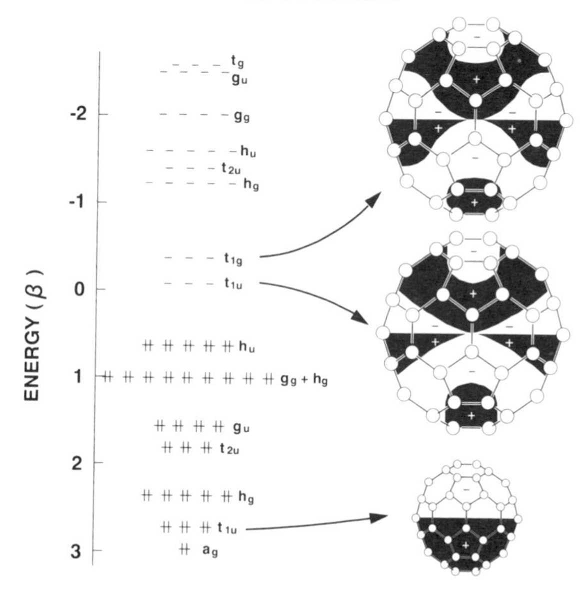

Figure 1. Hückel molecular orbital (HMO) energy levels of C_{60} together with one component of the triply degenerate t_{1u} and t_{1g} sets of molecular orbitals.

a result of the different solvents and reference systems that have been used. Of the measurements which are referenced to the standard calomel electrode, a representative first reduction potential for C_{60} is $E_{\frac{1}{2}} = -0.4V$ (SCE). There is also C^{13} NMR evidence for a highly reduced C_{60} species in THF solution in the presence of lithium metal which probably also corresponds to C_{60}^{6-} (Bausch *et al.* 1991). In the HMO diagram for C_{60} this reduction level corresponds to the complete filling of the t_{1u} level (figure 1). Other experiments, using lithium intercalation into C_{60}, have found evidence for even more highly reduced species, which would implicate the next set of triply degenerate levels in the HMO diagram (t_{1g}) (Chabre *et al.* 1992).

4. Solid state reduction

In the solid state, intercalation of C_{60} films, powders and crystals with alkali (A) metal dopants have led to new phases which may be insulators, conductors and superconductors depending on the filling of the t_{1u} level and the size of the dopant (Haddon *et al.* 1991; Hebard *et al.* 1991; Rosseinsky *et al.* 1991; Holczer *et al.* 1991; Tanigaki *et al.* 1991; Kochanski *et al.* 1992; Xiang *et al.* 1992; Fischer *et al.* 1992; Haddon 1992; Weaver 1992). For example, in the case of the A = K, Rb the A_3C_{60} phases are superconductors and the A_6C_{60} compounds are insulators. Thus when the t_{1u} level is half-filled (C_{60}^{3-}) the compounds are three-dimensional metals, but become insulators again when this level is full (C_{60}^{6-}). Recently it has been shown that the small alkali metals are able to produce A_xC_{60} phases with $x > 6$ (Fischer 1992).

Solid C_{60} also undergoes doping with alkaline earth (Ae) metals (Chen *et al.* 1992; Wertheim *et al.* 1992; Haddon *et al.* 1992) and Ca_5C_{60} has been shown to be a superconductor (Kortan *et al.* 1992). The nature of the doping with the alkaline earths is less clear than for the alkali metals. The alkali metal compounds appear to be fully charge transferred at least up to A_6C_{60} and exist as discrete phases. The alkaline earths, however, form solid solutions and the degree of charge transfer is less

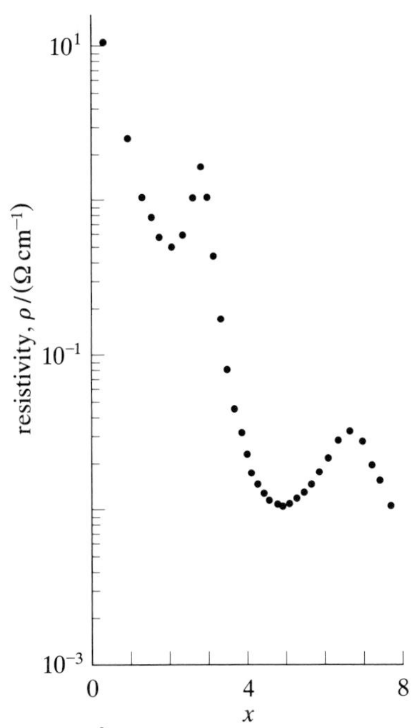

Figure 2. The resistivity of a 190 Å thick annealed Sr_xC_{60} film at 60° in UHV as a function of x (Haddon *et al.* 1992).

certain. Nevertheless, up to Ae_3C_{60}, photoemission (Chen *et al.* 1992; Wertheim *et al.* 1992), theoretical (Saito & Oshiyama 1992), and transport studies (Haddon *et al.* 1992) indicate a high degree of electron transfer and these phases correspond fairly well to C_{60}^{6-}. Beyond this level further doping apparently populates the t_{1g} level but the degree of charge transfer is uncertain for compositions in this range. For Ae_xC_{60} compositions with $x > 3$, photoemission studies suggest hybridization between the valence s-levels of the alkaline earths and the t_{1g} levels of C_{60}.

A plot of resistivity as a function of composition for a Sr_xC_{60} thin film (Haddon *et al.* 1992) is shown in figure 2 (similar behaviour is seen with Ca_xC_{60} films). There is a resistivity maximum close to the film composition Sr_3C_{60} and minima at Sr_2C_{60} and Sr_5C_{60}. The absence of a maximum at Sr_6C_{60} suggests that this composition is not fully charge transferred, and that hybridization with Sr states is now important in determining the electronic structure of the conducting phase. The nature of the doping beyond this point is even more uncertain, but the weak resistivity maximum at $x \approx 7$ in the Sr_xC_{60} thin film suggests that the t_{1g} band is effectively full at this point.

5. Theoretical interpretation within HMO theory

The basic structural features of the fullerenes have been recognized for some time. To close a sheet of carbon atoms into a fullerene, apart from six-membered rings (6MRs), spheroid formation requires the presence of 12 5MRs (Kroto *et al.* 1985; Zhang *et al.* 1986). The topological aspects of the fullerenes – placement of the 5MRs on the surface of the spheroid – is qualitatively accounted for in the simple HMO approach as shown in figure 1 for C_{60} (Haddon *et al.* 1986*a*). It may be seen that the two lowest unoccupied sets of molecular orbitals of C_{60} (t_{1u} and t_{1g}), bear a striking resemblance with respect to their nodal structure. To some extent this is exaggerated

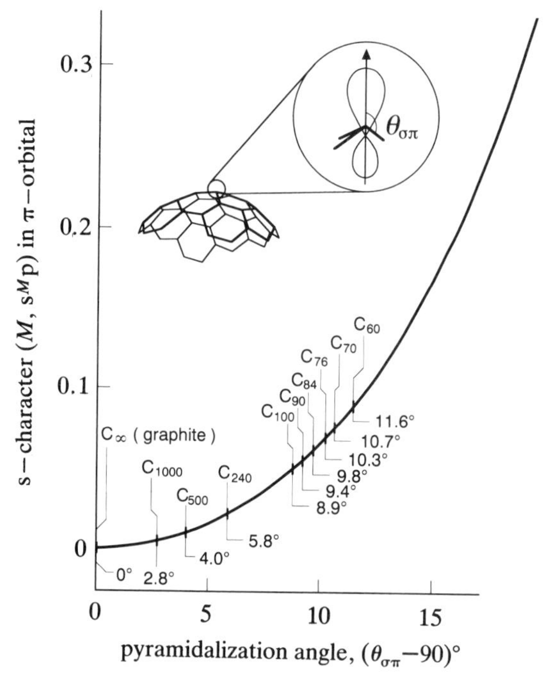

Figure 4. Rehybridization as a function of pyramidalization angle. The π-orbital axis vector (POAV1 approximation), is defined as that vector which makes equal angles to the three σ-bonds at a conjugated carbon atom (Haddon 1988). The common angle to the three σ-bonds (which are assumed to lie along the internuclear axes), is denoted $\theta_{\sigma\pi}$. The average pyramidalization angle $[(\overline{\theta}_{\sigma\pi}-90)°]$ shown for representative fullerenes (C_n), was obtained from eqn (2) of Haddon *et al.* (1986*b*) for $n > 60$.

situations. We have introduced the π-orbital axis vector (POAV) theory to analyse the electronic structure of nonplanar conjugated organic molecules such as the fullerenes (Haddon 1988). The POAV analysis extends σ–π separability into three-dimensions by use of the orbital orthogonality relations that are the basis of standard hybridization theory. In POAV2 theory, a π-orbital is defined as that hybrid orbital that is locally orthogonal to the σ-orbitals, whereas from the standpoint of the POAV1 analysis the π-orbital is that hybrid orbital which makes equal angles ($\theta_{\sigma\pi}$) with the σ-orbitals (figure 4).

Curvature in conjugated organic molecules leads the σ-bonds at the carbon atom to deviate from planarity and there is a change in hybridization: a rehybridization of the carbon atom so that π-orbital is no longer of purely p-orbital character. Thus the fullerenes are of intermediate hybridization (Haddon *et al.* 1986; Haddon & Raghavachari 1992). Using the standard nomenclature, the σ-bond hybridization falls between graphite (sp^2) and diamond (sp^3). Within certain limits the POAV theories allow a quantitative treatment of the hybridization in non-planar conjugated organic molecules and have been extensively tested (Haddon & Raghavachari 1992). The average POAV1 σ-bond hybridization for C_{60} is $sp^{2.278}$ and the π-orbital fractional s-character is 0.085 (POAV1) and 0.081 (POAV2) (Haddon 1990). The application of the POAV1 theory to the fullerenes takes a particularly simple form and an approximate treatment of the rehybridization required for closure of carbon spheroids of arbitrary size is shown in figure 4 where the curvature of the surface at a carbon atom is expressed by the pyramidalization angle $[(\theta_{\sigma\pi}-90)°]$. A recent analysis of ^{13}C NMR coupling constants in a derivative of C_{60} led to an estimate of 0.03 for the fractional rehybridization in neutral C_{60} (Hawkins *et al.* 1991; Hawkins 1992).

As may be seen in figure 4 and as noted above for C_{60} the fullerene π molecular orbitals are composed of carbon atomic orbitals which contain a substantial amount of 2s orbital character and as the carbon 2s orbital lies so much lower in energy than the 2p orbital the resulting orbitals are expected to exhibit an enhanced electronegativity when compared with their planar counterparts. Furthermore the fullerenes are not only of intermediate hybridization, but probably exhibit variable hybridization as they pass through different states of reduction, for the pyramidalization of anions is well known. Based on figure 3 it seems that the reduction potential of C_{60} is lowered by about 0.8V as a result of rehybridization effects.

The reductive behaviour of the higher fullerenes (C_n) should be of considerable interest as the 12 5MRs will ensure the presence of low lying energy levels. However, the rehybridization at each carbon atom decreases with increasing values of n and is asymmetrically distributed throughout the molecule for all fullerenes except C_{60} (Haddon & Raghavachari 1992). Furthermore the increasing size of these species will lead to a reduction in electron repulsion and make it easier to ion-pair the negatively charged surface.

7. Conclusion

C_{60} shows an extremely facile reduction profile and there is evidence for the addition of up to 12 electrons to the molecule. The prediction that C_{60} will exhibit an exceptionally high electron affinity and that the molecule will add up to 12 electrons under suitable conditions (Haddon *et al.* 1986*a*) seems to be borne out by the experimental results. Rehybridization plays an important role in determining the electronic structure of the fullerenes and it is the combination of topology and rehybridization which together account for the extraordinary ability of C_{60} to accept electrons.

References

Allemand, P.-M., Koch, A., Wudl, F., Rubin, Y., Diedrich, F., Alvarez, M. M., Anz, S. J. & Whetten, R. L. 1991 Two different fullerenes have the same cyclic voltammetry. *J. Am. chem. Soc.* **113**, 1050–1051.

Andersen Jr, A. G. & Masada, G. M. 1974 Polarographic reduction potentials of some nonbenzenoid aromatic hydrocarbons. *J. org. Chem.* **39**, 572–573.

Bausch, J. W., Prakash, G. K. S., Olah, G. A., Tse, D. S., Lorents, D. C., Bae, Y. K. & Malhotra, R. 1991 Diamagnetic polyanions of the C_{60} and C_{70} fullerenes: preparation, ^{13}C and ^{7}Li NMR spectroscopic observation, and alkylation with methyl iodide to polymethylated fullerenes. *J. Am. chem. Soc.* **113**, 3205–3206.

Breslow, R. & Grant, J. L. 1977 Electrochemical determination of the basicities of benzyl, allyl, and propargyl anions, and a study of solvent and electrolytic effects. *J. Am. chem. Soc.* **99**, 7745–7746.

Chabre, C., Djurado, D., Armand, M., Romanow, W. R., Coustel, N., McCauley Jr, J. P., Fischer, J. E. & Smith III, A. B. 1992 Electrochemical intercalation of lithium into solid C_{60}. *J. Am. chem. Soc.* **114**, 764–766.

Chen, E. C. M. & Wentworth, W. E. 1975 A comparison of experimental determinations of electron affinities of π-charge transfer complex acceptors. *J. chem. Phys.* **63**, 3183–3191.

Chen, Y., Stepniak, F., Weaver, J. H., Chibante, L. P. F. & Smalley, R. E. 1992 Fullerides of alkaline-earth metals. *Phys. Rev.* B **45**, 8845–8848.

Cox, D. M. *et al.* 1991 Characterization of C_{60} and C_{70} clusters. *J. Am. chem. Soc.* **113**, 2940–2944.

Curl, R. F. & Smalley, R. E. 1988 Probing C_{60}. *Science, Wash.* **242**, 1017–1022.

Dubois, D., Kadish, K. M., Flanagan, S., Haufler, R. E., Chibante, L. P. F. & Milson, L. J. 1991 Spectroelectrochemical study of the C_{60} and C_{70} fullerenes and their mono-, di-, tri-, and tetraanions. *J. Am. chem. Soc.* **113**, 4364–4366.

Dubois, D., Kadish, K. M., Flanagan, S. & Milson, L. J. 1991 Electrochemical detection of fulleronium and highly reduced fulleride (C_{60}^{5-}) ions in solution. *J. Am. chem. Soc.* **113**, 7773–7774.

Fischer, J. E., Heiney, P. A. & Smith III, A. B. 1992 Solid-state chemistry of fullerene-based materials. *Acc. Chem. Res.* **25**, 112–118.

Haddon, R. C. 1988 π-Electrons in three-dimensions. *Acc. Chem. Res.* **21**, 243–249.

Haddon, R. C. 1990 Measure of nonplanarity in conjugated organic molecules: Which structurally characterized molecule displays the highest degree of pyramidalization? *J. Am. chem. Soc.* **112**, 3385–3389.

Haddon, R. C. 1992 Electronic structure, conductivity, and superconductivity of alkali metal doped C_{60}. *Acc. Chem. Res.* **25**, 127–133.

Haddon, R. C. *et al.* 1991 Conducting films of C_{60} and C_{70} by alkali metal doping. *Nature, Lond.* **350**, 320–322.

Haddon, R. C., Brus, L. E. & Raghavachari, K. 1986*a* Electronic structure and bonding in icosahedral C_{60}. *Chem. Phys. Lett.* **125**, 459–464.

Haddon, R. C., Brus, L. E. & Raghavachari, K. 1986*b* Rehybridization and π-orbital alignment: the key to the existence of spheroidal carbon clusters. *Chem. Phys. Lett.* **131**, 165–169.

Haddon, R. C., Kochanski, G. P., Hebard, A. F., Fiory, A. T., Morris, R. C. 1992 Electrical resistivity and stoichiometry of Ca_xC_{60} and Sr_xC_{60} films. *Science, Wash.* **258**, 1636–1638.

Haddon, R. C. & Raghavachari, K. 1983 Naphthalene and azulene: theoretical comparison. *J. chem. Phys.* **79**, 1093–1094.

Haddon, R. C. & Raghavachari, K. 1992 Electronic structure of the fullerenes: carbon allotropes of intermediate hybridization, buckminsterfullerenes. *VCH*. (In the press.)

Haufler, R. E. *et al.* 1990 Efficient production of C_{60} (buckminsterfullerene), $C_{60}H_{36}$, and the solvated buckide ion. *J. phys. Chem.* **94**, 8634–8636.

Hawkins, J. M. 1992 Osmylation of C_{60}: proof and characterization of the soccer-ball framework. *Acc. Chem. Res.* **25**, 150–156.

Hawkins, J. M., Loren, S., Meyer, A. & Nunlist, R. 1991 2D nuclear magnetic resonance analysis of osmylated C_{60}. *J. Am. chem. Soc.* **113**, 7770–7771.

Hebard, A. F., Rosseinsky, M. J., Haddon, R. C., Murphy, D. W., Glarum, S. H., Palstra, T. T. M., Ramirez, A. P. & Kortan, A. R. 1991 Superconductivity at 18 K in potassium-doped C_{60}. *Nature, Lond.* **350**, 600–601.

Holczer, K., Klein, O., Huang, S.-M., Kaner, R. B., Fu, K.-J., Whetten, R. L. & Diederich, F. 1991 Alkali-fulleride superconductors: synthesis, composition, and diamagnetic shielding. *Science, Wash.* **252**, 1154–1157.

Kochanski, G. P., Hebard, A. F., Haddon, R. C. & Fiory, A. T. 1992 Electrical resistivity and stoichiometry of K_xC_{60} films. *Science, Wash.* **255**, 184–186.

Kortan, A. R., Kopylov, N., Glarum, S. H., Gyorgy, E. M., Ramirez, A. P., Fleming, R. M., Thiel, F. A. & Haddon, R. C. 1992 Superconductivity at 8.4 K in calcium-doped C_{60}. *Nature, Lond.* **355**, 529–532.

Krätschmer, W., Lamb, L. D., Fostiropoulos, K. & Huffman, D. R. 1990 Solid C_{60}: a new form of carbon. *Nature, Lond.* **347**, 354–358.

Kroto, H. W., Heath, J. R., O'Brien, S. C., Curl, R. F. & Smalley, R. E. 1985 C_{60}: Buckminsterfullerene. *Nature, Lond.* **318**, 162–164.

Limbach, P. A., Schweikhard, L., Cowen, K. A., McDermott, M. T., Marshall, A. G. & Coe. J. V. 1991 Observation of the doubly charged, gas-phase fullerene anions C_{60}^{2-} and C_{70}^{2-}. *J. Am. chem. Soc.* **113**, 6795–6798.

Miller, B., Rosamilia, J. M., Dabbagh, G., Muller, A. J. & Haddon, R. C. 1992 Electron transfer to C_{60} and C_{70} fullerenes at hydrodynamic and dual electrodes. *J. electrochem. Soc.* **139**, 1941–1945.

Rosseinsky, M. J. *et al.* 1991 Superconductivity at 28 K in Rb_xC_{60}. *Phys. Rev. Lett.* **66**, 2830–2832.

Saito, S. & Oshiyama, A. 1992 Electronic structure of calcium-doped C_{60}. *Solid State Commun.* **83**, 107–110.

Streitwieser 1962 *Molecular Orbital Theory for organic chemists*, ch. 7. New York: Wiley.

Streitwieser Jr, A. & Schwager, I. 1962 A molecular orbital study of the polarographic reduction in dimethylformamide of unsubstituted and methyl-substituted aromatic hydrocarbons. *J. Am. chem. Soc.* **66**, 2136–2320.

Tanigaki, K., Ebbesen, T. W., Saito, S., Mizuki, J., Tsai, J. S., Kubo, Y. & Kuroshima, S. 1991 Superconductivity at 33 K in $Cs_xRb_yC_{60}$. *Nature, Lond.* **352**, 222–223.

Trost, B. M., Bright, G. M., Frihart, C. & Britelli, D. 1971 Perturbed [12]annulenes. The synthesis of pyracylenes. *J. Am. chem. Soc.* **93**, 737–745.

Wang, L. S., Conceicao, J., Jin, C. & Smalley, R. E. 1991 Threshold photodetachment of cold C_{60}^-. *Chem. Phys. Lett.* **182**, 5–11.

Weaver, J. H. 1992 Fullerenes and fullerides: photoemission and scanning tunneling microscopy studies. *Acc. Chem. Res.* **25**, 143–149.

Wertheim, G. K., Buchanan, D. N. E. & Rowe, J. E. 1992 Charge donation by Ca into the t_{1g} band of C_{60}. *Science, Wash.* **258**, 1638–1640.

Wudl, F. 1992 The chemical properties of buckminsterfullerene (C_{60}) and the birth and infancy of fulleroids. *Acc. Chem. Res.* **25**, 157–161.

Xiang, X.-D., Hou, J. G., Briceno, G., Vareka, W. A., Mostovoy, R., Zettl, A., Crespi, V. H. & Cohen, M. L. 1992 Synthesis and electronic transport of single crystal K_3C_{60}. *Science, Wash.* **256**, 1190–1191.

Xie, Q., Perez-Cordero, E. & Echegoyen, L. 1992 Electrochemical detection of C_{60}^{6-} and C_{70}^{6-}: Enhanced stability of fullerides in solution. *J. Am. chem. Soc.* **114**, 3978–3980.

Yildirim, T., Zhou, O., Fischer, J. E., Bykovetz, N., Strongin, R. A., Cichy, M. A., Smith III, A. B., Lin, C. L. & Jelivek, R. 1992 Intercalation of sodium heteroclusters into the C_{60} lattice. *Nature, Lond.*, **360**, 568–571.

Zhang, Q. L., O'Brien, S. C., Heath, J. R., Liu, Y., Curl, R. F., Kroto, H. W. & Smalley, R. E. 1986 Reactivity of large carbon clusters: spheroidal carbon shells and their possible relevance to the formation and morphology of soot. *J. phys. Chem.* **90**, 525–528.

Discussion

R. E. Palmer (*University of Cambridge, U.K.*). If, hypothetically, you could continue to stretch the spacing in the superconductor, there must come a point where they become insulating again. Is there some theoretical maximum for the T_c that you can envisage?

R. C. Haddon. Cs_3 may represent that point. It seems by the time you get to Cs_3 the spacing between the balls that is necessary to accommodate Cs becomes too large. I don't know if it is an argument for the electronic structure of the material, although it always seems that superconductors are on the verge of some sort of instability. I don't know of any theoretical maximum in terms of the T_c, but I think as you start to move the balls apart, the band width will get so narrow that localization will develop.

H. W. Kroto (*University of Sussex, U.K.*). Is there anything special about the number 3, or does no one know? Should one be looking for another fullerene which has maybe, double degeneracy?

R. C. Haddon. I think there is something special. We do need more fullerenes with triply degenerate orbitals; unfortunately none have been isolated beside C_{60}. It comes about as a result of the direct product of the t-type orbitals.

P. A. SERMON (*Brunel University, U.K.*). Does your doping procedure only involve alkali metals or is there oxygen involved?

R. C. HADDON. We don't believe that there is any oxygen involved.

P. A. SERMON. May I take it that this type of superconductivity is quite different from that of the ceramics that have been described in the literature?

R. C. HADDON. I think so. The explanation that I have mentioned seems to be fairly successful in accounting for the result. It is based on vibrations, even though they are molecular vibrations. As far as I know, there is no accepted theory of superconductivity in ceramics. In fact, that is one of the uses we can point to for C_{60} superconductors. I think that, in the end, these will be among the best understood of the superconductors.

The carbon-bearing material in the outflows from luminous carbon-rich stars

BY M. JURA
Department of Astronomy, University of California, Los Angeles, California 90024, U.S.A.

Within the neighbourhood of the Sun, a number of highly evolved stars are carbon-rich in the sense that they have more carbon than oxygen so their outer atmospheres contain molecules such as CN, CH and C_2H_2. These stars are cool with atmospheric temperatures near 3000 K and they are also luminous, typically 10^4 times more powerful than the Sun. The outer envelopes of these stars are tenuously bound, and they all are losing mass at a very high rate, in some cases more than $10^{-5}\,M_\odot\,a^{-1}$ (where $M_\odot$ denotes the mass of the Sun). These high luminosity carbon stars remain in this phase for a time, very approximately, near 10^5 years. They exhibit a large amount of carbon in their atmospheres because the products of the nuclear burning that occurs in the very centre of the star, including the synthesis of carbon, appear on the surface.

In the extended envelopes around these stars, there is a very active chemistry, and the gas is sufficiently cool that nucleation of solid dust grains occurs. These solid particles may grow to sizes as large as 1 μm although a more typical size is near 0.05 μm. We therefore can identify both relatively small carbon-bearing molecules (for example HC_7N) and much larger carbon-containing dust grains in the outflows. The amount of intermediate size particles or molecules, such as C_{60}, and their possible role in the circumstellar chemistry is not yet well understood. At least in the envelope of the well studied carbon star IRC + 10216, there appears to be more carbon in CO and solid grains than in polycyclic aromatic hydrocarbons.

1. Introduction

As stars evolve, they expand into luminous, red giants (see, for example, Iben & Renzini 1983). When the red giant becomes sufficiently luminous, perhaps $L > 2000\,L_\odot$, where $L_\odot$ denotes the luminosity of the Sun, the mass loss rate becomes detectably large (more than $10^{-7}\,M_\odot\,a^{-1}$). Indeed, mass loss rates approaching $10^{-4}\,M_\odot\,a^{-1}$ are known (Jura 1991*b*).

Infrared and radio observations of the circumstellar envelope have been exploited very successfully. The most commonly studied molecule is CO; hundreds of nearby stars are known CO sources (Nyman *et al.* 1992). Additionally, over 50 different species have now been detected in various circumstellar envelopes (Olofsson 1992).

A star whose initial mass is between 1 and about $5\,M_\odot$ (an 'intermediate mass' star) is thought to evolve into a white dwarf of typically 0.6 to $1\,M_\odot$ (Weidemann & Koester 1983). As a result, main sequence stars can lose more than 50% of their initial mass. Since the Chandrasekhar limit for the maximum mass of a white dwarf is near $1.4\,M_\odot$, if this mass loss did not occur, there would be many more supernova and many fewer white dwarfs in the Milky Way than are observed.

Phil. Trans. R. Soc. Lond. A (1993) **343**, 63–72

Printed in Great Britain

Both because the amount of mass that is lost is so substantial, and because there can be mixing of material from the very centre into outer layers of the star, the composition of the surface of a star losing a great deal of mass may reflect the large amount of nucleosynthesis that has occurred during the evolution of the star. Highly evolved mass-losing stars may display a number of distinctive elemental abundances. About half of the intermediate mass stars that lose mass very rapidly (that is $dM/dt > 2 \times 10^{-6} M_{\odot}\ a^{-1}$) are carbon-rich (Jura & Kleinmann 1989). Therefore, carbon-rich environments are common.

In the atmosphere of a red giant star, the most stable molecule is CO. By standard thermodynamic arguments, most of the oxygen and carbon is contained within this molecule. If the star is oxygen-rich, then [O] > [C] and the atmosphere contains molecules such as H_2O as well as CO. If the star is carbon-rich, then [C] > [O] and the atmosphere contains large amounts of molecules such as CN and CH (Wallerstein 1973; Blanco 1989). Stars with [C] ≈ [O] are denoted as S-type; there are about a third as many S-type stars as there are carbon stars (Jura 1988). In the general galactic population carbon stars are very rare; however, as noted above, in the population of stars returning large amounts of mass to the interstellar medium, the carbon stars are common.

The carbon-rich stars losing a large amount of mass are a fascinating laboratory for studying carbon chemistry in space. Below, I describe what is known about these stars, and how we can hope to use them to understand the nature of celestial carbon-chemistry.

2. Circumstellar molecules

The best studied mass-losing carbon star is IRC+10216 which appears to be only 130 pc from the Sun; the closest of all these objects (see Jura 1991*b*). This star is probably losing mass at a rate, dM/dt, of about $2 \times 10^{-5} M_{\odot}\ a^{-1}$ while the gas is flowing out at a speed, v_{∞}, of about 15 km s^{-1}. As a first approximation it seems that at least for many of these stars, we may assume a spherically symmetric mass loss. Consequently, we may write for the density, ρ, as a function of distance from the star, R, that

$$\rho = (dM/dt)/(4\pi v_{\infty} R^2). \tag{1}$$

The circumstellar chemistry is often subdivided into three main zones, which are determined by a comparison of the characteristic dynamic flow time, R/v_{∞}, with the chemical reaction times (Lafont *et al.* 1982; Omont 1987; Millar 1988). (i) In the region closest to the star (perhaps $R \approx 10^{14}$ cm), the density is sufficiently high that three-body chemical reactions occur in a time short compared to the dynamic time. In this régime, we expect the chemical abundances to approach thermodynamic equilibrium. (ii) Somewhat further away from the star (10^{14} cm $< R < 10^{16}$ cm), there is a 'freeze-out' of the products of the three-body reactions (McCabe *et al.* 1979). In this region, two-body reactions dominate the active chemistry. (iii) Finally, far from the star ($R > 10^{16}$ cm), the density becomes sufficiently low that the only significant chemical processing is the photodestruction that results from absorption of ambient interstellar ultraviolet photons by the resulting molecules that flow from the central star.

For carbon-rich outflows, this scheme seems to predict, reasonably well, the observed abundances and spatial distributions of some of the most abundant circumstellar molecules such as CO and HCN (Olofsson *et al.* 1990; Cherchneff &

Barker 1992). Values of $[HCN]/[H_2]$ as large as 10^{-4} are known (Jura 1991*a*), although this inferred ratio is very high compared with the value for most carbon stars (e.g. $[HCN]/[H_2] \approx 10^{-6}$ toward IRC+10216 according to Bieging *et al.* (1984)). Because the derived values of $[HCN]/[H_2]$ are substantially lower than the estimated values for $[CO]/[H_2]$ of 8×10^{-4} (Knapp & Morris 1985), it seems, as predicted by the models, that there is much more carbon contained within CO than within HCN. While detailed schemes to account for a number of other species in the outer envelope also have been successful (Glassgold *et al.* 1987; Nejad & Millar 1987), the formation of the larger species, such as HC_7N, is still unclear (see Jura & Kroto 1990).

It has not been possible to determine directly the carbon abundance ([C]/[H]) in the atmosphere of a carbon star losing a very large amount of mass (*ca.* $10^{-5}\,M_{\odot}\,a^{-1}$). However, such stars are thought to expel their envelopes and evolve into planetary nebulae in less than 10^5 years, and it is possible to measure directly the gas-phase abundances of carbon ([C]/[H]) and oxygen ([O]/[H]) in planetaries. Zuckerman & Aller (1986) report [C]/[O] abundances for 29 carbon-rich planetary nebulae, the likely evolutionary descendants of mass-losing carbon-rich stars. For these 29 planetaries, the average value of [C]/[O] is 2.3. Therefore, under the plausible assumption that while the star is a mass-losing red giant, essentially all the oxygen is contained within CO, there is still a lot of 'leftover' carbon, and it is quite possible that there is considerable additional carbon beyond what is found in CO or the solid dust grains.

One possible important reservoir of circumstellar carbon is C_2H_2. According to the theoretical models presented by Lafont *et al.* (1982), there may be 0.6 as much carbon being carried in C_2H_2 as there is in CO around IRC+10216. However, Keady & Hinkle (1988) have observed infrared line strengths to infer that $[C_2H_2]/[H_2] = 5 \times 10^{-5}$. Therefore, according to their results, there is perhaps 0.1 as much carbon in C_2H_2 as in CO.

There is additional evidence that the abundance of C_2H_2 is relatively low in the outflow from IRC+10216. According to the standard models, we expect that in the outer envelope to IRC+10216, the C_2H_2 is photodissociated to produce C_2H (Glassgold *et al.* 1986). That is, the reaction

$$h\nu + C_2H_2 \rightarrow C_2H + H, \quad (2)$$

should produce large concentrations of C_2H. Neither from infrared (Keady & Hinkle 1988) nor radio observations (Truong-Bach *et al.* 1987) does there seem to be as much C_2H as predicted by the photochemical models if C_2H_2 is abundant as 10^{-4} H_2. Because C_2H_2 might be a key molecule in the synthesis of larger molecules (Keller 1987), it is possible that there is relatively little C_2H_2 because it is consumed in the production of larger species.

3. Solid grains

Evidence for solid grains in the mass loss around carbon stars comes from (i) the infrared continuum emission which is far in excess of what is expected from the photosphere of the star. This excess radiation is thought to be the emission from relatively cool circumstellar dust grains. (ii) Many carbon-rich stars display strong emission near 11.3 μm thought to be the result of solid carbon (Papoular 1988). (iii) Some carbon stars exhibit reflection nebulosity which is the consequence of scattering by circumstellar dust grains (see, for example, Tamura *et al.* 1988). (iv) Often, a carbon star will display a net polarization because the circumstellar dust grains are distributed asymmetrically with the consequence of producing a net scattering.

While there is considerable uncertainty in the size distribution and composition of circumstellar dust grains, much of the evidence points towards some sort of amorphous carbon with a range of sizes up to *ca.* 1 μm although it seems that the grains can be successfully modelled with particles of radius 0.05 μm (Martin & Rogers 1987). Therefore, we anticipate that there are solid particles containing upwards of *ca.* 10^{11} atoms.

Jura (1986) has argued that for the carbon stars losing large amounts of mass on average, the mass of carbon in dust grains relative to the total amount of hydrogen is 4.5×10^{-3}, if the opacity at 60 μm is 150 $cm^2\ g^{-1}$. Because this dust to gas ratio is derived with the assumption that $[CO]/[H_2] = 8 \times 10^{-4}$, this result for the dust abundance implies that there is, on average, twice as much carbon contained within CO as there is contained within the solid grains. These data would therefore imply that for the stars losing a large amount of mass that [C]/[O] = 1.3. This result depends upon the infrared opacity of the circumstellar grains which is not well known. Nevertheless, if this model is correct and if, on average, [C]/[O] in the circumstellar envelopes of these stars in fact equals 2.3, then it appears that there should be some additional reservoir of carbon in the outflows from carbon rich stars besides CO and solid grains.

The nature of this other carrier of the carbon is unknown. It is possible that these other forms of carbon-bearing species are relatively fragile in the sense that they may survive the relatively cool and protected wind of the red giant and are then destroyed in the harsh environments of Planetary Nebulae that follow. Two arguments suggest that substantial erosion of carbon-bearing material occurs in carbon-rich planetary nebulae. (i) The [C]/[O] ratio in the ionized regions in carbon-rich planetary nebula averages 2.3, therefore there must be an appreciable source of gas-phase carbon beyond the photodestruction of the CO molecule. (ii) The inferred dust to gas ratio in the ionized regions of old, extended planetaries is between 10^{-3} and 10^{-4} (Pottasch *et al.* 1984), substantially less than the dust to gas ratio of 4.5×10^{-3} inferred for the red giant progenitors to these stars. Unless the analysis of the infrared data is substantially in error (see, for example, Huggins & Healy 1989), there has been substantial destruction of the carbon-bearing particles (grains, clusters, molecules) as the object evolved from being a red giant with an extended circumstellar envelope into a planetary nebula.

4. Intermediate size clusters

While we have direct observational evidence for small molecules and large grains, we have very little information about clusters in the outflows from carbon-rich stars. That is, we know very little about the particles with more than 10 atoms, but, say, fewer than 10^6 atoms in circumstellar regions (see Kroto & Jura 1992).

Infrared spectroscopy has revealed the presence of the 'Unidentified features' in carbon-rich planetary nebulae and other sources. There are particularly prominent features at 3.3, 6.2, 7.7, 8.6 and 11.3 μm (Sellgren 1990). These features are not detected in mass-losing carbon stars, but they are strong in 'transition objects' that are evolving from red giants to planetary nebulae such as the Egg Nebula, AFGL 2688 (Geballe *et al.* 1992), and the Red Rectangle (Geballe *et al.* 1989).

The infrared features are generally attributed to various vibrational modes of carbonaceous material, quite possibly PAHs (polycyclic aromatic hydrocarbons: Léger & Puget 1984; Allamandola *et al.* 1985) or HAC (hydrogenated amorphous carbon: Duley & Williams 1981). However, while the presence of the emission features provides evidence for organic compounds, it has not been possible to make specific identifications. Also, there are significant differences in the infrared spectral features in different carbon-rich objects. For example, the profile of the feature at 3.29 μm toward the Red Rectangle is quite different from that toward the carbon-rich planetary nebula NGC 7027 (Tokunaga *et al.* 1988).

It should also be noted that the PAH features appear to be primarily excited in regions where there is a substantial energy density of ultraviolet photons are present but where the hydrogen is neutral. That is, in the Orion Bar region, there is a strong anti-correlation between regions where the 3.3 μm feature is strong and where the Brackett α recombination line of ionized hydrogen is strong (Sellgren *et al.* 1990). Similarly, in the carbon-rich planetary nebulae, NGC 7027, the emission from the dust features appears to be more extended than the emission from the ionized gas (Aitken & Roche 1983; Woodward *et al.* 1989).

To date, neither PAH emission nor absorption has been detected in the circumstellar envelope around a cool carbon star; PAH emission has only been seen in carbon-rich environments where there is substantial energy density of ultraviolet radiation. This correlation could simply be an excitation effect; the carbon features are only excited by the presence of ultraviolet radiation. However, it could also be that carbon particles are eroded into PAHs in the environment where ultraviolet penetrates; either directly by the ultraviolet radiation or indirectly by shocks that accompany the radiation.

Here we place an upper limit to the amount of PAHs flowing out of the very well studied carbon star, IRC+10216 by placing limits on the strength of the PAH feature at 3.3 μm. Because the 3.3 μm feature is intrinsically quite broad, it is useful to consider relatively low resolution observations. Witteborn *et al.* (1980) have presented a spectrum with 2% resolution between 2.0 and 8.5 μm of IRC+10216. Treffers & Cohen (1974) and Merrill & Stein (1976) have presented similar data over much of the same spectral interval. Neither group detects any absorption at 3.28 μm at more than 5% of the continuum level. There is notable absorption at 3.1 μm, but this 'feature' is believed to be a blend of sharp molecular lines (Ridgway *et al.* 1978).

We can estimate the column density of material in a circumstellar shell, N_H, extending from radius R to infinite distance, from the expression:

$$N_H = (\mathrm{d}M/\mathrm{d}t)/(4\pi \mu R v_\infty), \tag{3}$$

where μ is the atomic weight of hydrogen. For IRC+10216, we adopt a distance of 130 pc (Jura 1991*b*), a mass loss rate of $2 \times 10^{-5}\ M_\odot\ \mathrm{a}^{-1}$ (Kwan & Linke 1982; Martin & Rogers 1987) with an outflow speed of 15 km s^{-1}. We also adopt an inner radius of 10^{15} cm from which the PAHs might be measured (see Keady *et al.* 1988). From these numbers we find that $N_H = 4.0 \times 10^{22}\ \mathrm{cm}^{-2}$. We adopt for the 3.28 μm PAH feature that $\sigma_{\text{C-H}} = 3.5 \times 10^{-20}\ \mathrm{cm}^{-2}$ (Léger *et al.* 1989) for each C–H bond in a PAH. The absence of a 3.3 μm feature at the 5% level therefore implies that $N_{\text{C-H}} < 1.4 \times 10^{18}\ \mathrm{cm}^{-2}$. Therefore, combining the two results, we may write that

$$N_{\text{C-H}}/N_{\mathrm{H}} < 3.5 \times 10^{-5}. \tag{4}$$

If we assume that about a third of all the carbon atoms within a PAH have an associated hydrogen atom, then we find that the amount of carbon within PAHs flowing out of IRC+10216 is less than about 10^{-4} of the hydrogen nuclei. Since $[CO]/[H_2]$ is about 8×10^{-4}, it seems that there is more carbon in CO and in grains than in PAHs flowing out of IRC+10216. Consequently, PAHs, by themselves, do not carry most of any 'excess' carbon, the carbon beyond that in CO and grains, in the outflow from IRC+10216.

Even though there may not be especially large abundances of PAHs flowing out of carbon stars such as IRC+10216, the possibility remains that the putative clusters of carbon particles flowing out of carbon-rich red giants may be related to a fascinating unsolved problem in astrophysics; the origin of the diffuse interstellar bands (Herbig 1975). These diffuse bands are found in absorption throughout the interstellar medium, but we do not know their carrier. Most carbon stars display such complex spectra that it is very difficult to search for the diffuse bands in their spectra. Pritchett & Grillmair (1984) reported that the diffuse band at 5780 Å† was particularly strong in the absorption spectrum of the carbon-rich planetary nebula NGC 7027. However, this object lies close to the galactic plane, and it is not certain whether the absorption features have a circumstellar or interstellar origin. Le Bertre & Lequeux (1992) argue that there is no evidence for an enhancement of the diffuse bands in the circumstellar matter around NGC 7027.

Le Bertre (1990) has found that the features at 4430, 5780 and 6284 Å are quite strong in the absorption spectrum of the A star companion to the mass-losing carbon star CS 776 (= IRC-20131). The diffuse band at 5797 Å is not present in the spectrum of this companion. Again, however, because the star lies in the galactic plane ($b = -0.81°$), much of the diffuse bands may be contributed by interstellar instead of circumstellar matter.

This difficulty for CS 776 can be assessed quantitatively. Le Bertre (1990) derives a mass loss rate from IRC-20131 of $4.5\times10^{-7}\,M_{\odot}\,a^{-1}$. Scaled to the outflow velocity of the material of 26 km s^{-1} (Zuckerman & Dyck 1989) instead of the assumed value of 15 km s^{-1} and using Le Bertre's distance of 1.3 kpc instead of their estimate of 1.43 kpc, the re-computed mass loss rate from Claussen *et al.* (1987) is $6.6\times10^{-7}\,M_{\odot}\,a^{-1}$ in reasonable agreement with the rate estimated by Le Bertre (1990). Because the separation of the A star companion from the carbon star CS 776 is 1.81″, the projected separation between the two stars is 3.6×10^{16} cm. Therefore, according to equation (3), the column density of circumstellar hydrogen between us and the A star companion to CS 776 is 1.5×10^{19} cm^{-2}. However, the total extinction toward this companion is $A_V = 1.71$ mag (Le Bertre 1990) which, for a standard interstellar dust to gas ratio, corresponds to a hydrogen column density of 3×10^{21} cm^{-2} (Spitzer 1978). This column density is consistent with the expected concentration of interstellar matter within the plane of the Milky Way. Thus, towards the companion to CS 776, there appears to be about 100 times more interstellar than circumstellar matter. Therefore, unless the diffuse bands are extremely strong in the circumstellar matter around CS 776, it seems quite likely that the bulk of the diffuse bands in the spectrum result from interstellar matter.

The best evidence for a relation between carbon-particles and the diffuse interstellar bands comes from analysis of the Red Rectangle. The Red Rectangle is an usual mass-losing carbon star which is probably in transition into becoming a planetary nebula. Schmidt *et al.* (1980) using 6–20 Å resolution discovered intense optical emission bands longward of 5400 Å. With a higher spectral resolution of 1 Å,

† 1 Å = 10^{-10} m = 10^{-1} nm.

Warren-Smith *et al.* (1991) resolved individual lines in this broad emission band. Sarre (1991) and Fossey (1990) have independently pointed out that the emission lines at 5799, 5855, 6380 and 6615 Å discovered in this high resolution spectrum of the Red Rectangle agree in wavelength very well with some of the strongest interstellar diffuse absorption bands that lie at 5797, 5850, 6376, 6379 and 6614 Å (Herbig 1975). It seems likely, therefore, that the carrier of at least some of the diffuse interstellar bands is being produced in the outflow from the Red Rectangle. In particular, Scarrott *et al.* (1992) have shown that the line profile of the bands both in emission and absorption can be reproduced by a complex molecules such as 'C_{60}-entity' (e.g. C_{60}^+ or $C_{60}X$). Other complex carbon molecules such as PAHs and HAC have also been proposed to be the carrier of the broad bands in the Red Rectangle (d'Hendecourt *et al.* 1986; Wdowiak *et al.* 1989; Duley & Williams 1990).

Another possibility is that the diffuse bands are carried by PAHs. Recently, Salama & Allamandola (1992) have shown that $C_{16}H_{10}^+$ in an argon matrix has a strong absorption feature at 4435 Å, close to the strongest diffuse interstellar band at 4430 Å. However, in a neon matrix in which molecule-matrix interactions are expected to be less severe, the same absorption occurs at 4395 Å. Therefore, it is still uncertain whether gas-phase $C_{16}H_{10}^+$ exists in large enough quantities in the interstellar medium to produce the feature and whether this particular gas-phase ion absorbs at the precise wavelength to account for the diffuse interstellar band. Nevertheless, it is quite possible that any $C_{16}H_{10}$ in the circumstellar envelope around the Red Rectangle is ionized; Balm & Jura (1992) and Hall *et al.* (1992) have identified CH^+ in the outflow from this star from the spectra of Waelkens *et al.* (1992).

With current instruments it is possible to make spatial maps of the emission from different species in the Red Rectangle. These maps might provide valuable clues to the origin of different spectroscopic features. For example, in the spectrum of the Red Rectangle, the emission features which correspond to the diffuse interstellar bands are concentrated in what appears to be two hollow cones oriented perpendicular to the plane of this bipolar system (Schmidt & Witt 1991). This hollow cone is similar to that proposed by Jura & Kroto (1990) to explain the observed (Nguyen-Q-Rieu *et al.* 1986) HC_7N emission (see around AFGL 2688, the 'Egg Nebula'), a very well studied carbon-rich object that appears to be in transition from a red giant to a planetary nebula.

5. Conclusions

Mass-losing carbon stars are major sources of carbon-rich material in the interstellar medium. In addition to small carbon-bearing molecules and large carbon grains, there may be a substantial amount of intermediate-size carbon clusters (C_{60}, PAHs, etc.) flowing out of these objects. At least towards IRC+10216, there appears to be more carbon in CO and in grains than in PAHs. A promising possibility is that the diffuse interstellar bands may be carried by carbon-rich clusters and that the identification of the carrier of the bands may also prove valuable insight into the nature of these intermediate-size carbon clusters in astrophysics.

I thank Simon Balm and Harry Kroto for many useful comments. This work has been partly supported by NASA.

References

Aitken, D. K. & Roche, P. R. 1983 *Mon. Not. R. astronom. Soc.* **202**, 1233.

Allamandola, L. J., Tielens, A. G. G. M. & Barker, J. R. 1985 *Astrophys. J.* **290**, L25.

Balm, S. P. & Jura, M. 1992 *Astronom. Astrophys.* **261**, L25.

Bieging, J. H., Chapman, B. & Welch, W. J. 1984 *Astrophys. J.* **285**, 656.

Blanco, V. M. 1989 *Revista Mexicana Astronomia Astrofisica* **19**, 25.

Cherchneff, I. & Barker, J. R. 1992 *Astrophys. J.* **394**, 703.

d'Hendecourt, L. B., Léger, A., Olofsson, G. & Schmidt, W. 1986 *Astronom. Astrophys.* **170**, 91.

Duley, W. W. & Williams, D. A. 1981 *Mon. Not. R. astronom. Soc.* **196**, 269.

Duley, W. W. & Williams, D. A. 1990 *Mon. Not. R. astronom. Soc.* **247**, 147.

Fossey, S. F. 1990 Ph.D. thesis, University College, London, U.K.

Geballe, T. R., Tielens, A. G. G. M., Allamandola, L. J., Moorhouse, A. & Brand, P. W. J. L. 1989 *Astrophys. J.* **341**, 278.

Geballe, T. R., Tielens, A. G. G. M., Kwok, S. & Hrivnak, B. J. 1992 *Astrophys. J.* **387**, L89.

Glassgold, A. E., Lucas, R. & Omont, A. 1986 *Astronom. Astrophys.* **157**, 35.

Glassgold, A. E., Mamon, G. A., Omont, A. & Lucas, R. 1987 *Astronom. Astrophys.* **180**, 183.

Hall, D. I., Miles, J. R., Sarre, P. J. & Fossey, S. J. 1992 *Nature Lond.* **358**, 629.

Herbig, G. H. 1975 *Atrophys. J.* **196**, 129.

Huggins, P. J. & Healy, A. P. 1989 *Astrophys. J.* **346**, 201.

Iben, I. & Renzini, A. 1983 *A. Rev. Astronom. Astrophys.* **21**, 271.

Jura, M. 1986 *Astrophys. J.* **303**, 327.

Jura, M. 1988 *Astrophys. J. (suppl.)* **66**, 33.

Jura, M. 1991*a* *Astrophys. J.* **372**, 208.

Jura, M. 1991*b* *Astronom. Astrophys. Res.* **2**, 227.

Jura, M. & Kleinmann, S. G. 1989 *Astrophys. J.* **341**, 359.

Jura, M. & Kroto, H. 1990 *Astrophys. J.* **351**, 222.

Keller, R. 1987 In *Polycyclic aromatic hydrocarbons and astrophysics* (ed. A. Léger, L. d'Hendecourt & N. Boccara). Dordrecht: Reidel.

Keady, J. J., Hall, D. N. B. & Ridgway, S. T. 1988 *Astrophys. J.* **326**, 832.

Keady, J. J. & Hinkle, K. H. 1988 *Astrophys. J.* **331**, 539.

Knapp, G. R. & Morris, M. 1985 *Astrophys. J.* **292**, 640.

Kroto, H. & Jura, M. 1992 *Astronom. Astrophys.* **263**, 275

Lafont, S., Lucas, R. & Omont, A. 1982 *Astronom. Astrophys.* **106**, 201.

Léger, A., d'Hendecourt, L. & Défourneau, D. 1989 *Astronom. Astrophys.* **216**, 148.

Léger, A. & Puget, J.-L. 1984 *Astronom. Astrophys.* **137**, L5.

Le Bertre, T. 1990 *Astronom. Astrophys.* **236**, 472.

Le Bertre, T. & Lequeux, J. 1992 *Astronom. Astrophys.* **255**, 288.

Martin, P. G. & Rogers, C. 1987 *Astrophys. J.* **322**, 374.

McCabe, E. M., Smith, R. C. & Clegg, R. E. S. 1979 *Nature Lond.* **281**, 263.

Merrill, K. M. & Stein, W. A. 1976 *Proc. astronom. Soc. Pacific* **88**, 294.

Millar, T. J. 1988 In *Rate coefficients in astrochemistry* (ed. T. J. Millar & D. A. Williams), p. 287. Reidel: Dordrecht.

Nejad, L. A. M. & Millar, T. J. 1987 *Astronom. Astrophys.* **183**, 279.

Nguyen-Q-Rieu, Winnberg, A. & Bujarrabal, V. 1986 *Astrophys. J.* **165**, 204.

Nyman, L.-Å., Booth, R. S., Carlstrom, U., Habing, H. J., Heske, A., Sahai, R., Stark, R., van der Veen, W. E. C. J. & Winnberg, A. 1992 *Astronom. Astrophys. (Suppl.)* **93**, 121.

Olofsson, H. 1992 In *Proc. ESO/CTIO Workshop on Mass Loss on the AGB and Beyond*. La Serena, Chile. (In the press.)

Olofsson, H., Eriksson, K. & Gustafsson, B. 1990 *Astronom. Astrophys.* **230**, 405.

Omont, A. 1987 In *IAU Symp. no. 120 on Astrochemistry* (ed. M. S. Vardya & S. P. Tarafdar), p. 357. Dordrecht: Reidel.

Papoular, R. 1988 *Astronom. Astrophys.* **204**, 138.

Pottasch, S. R., Baud, B., Beintema, D., Emerson, J., Habing, H. J., Houck, J., Jennings, R. & Marsden, P. 1984 *Astronom.* **138**, 10.

Pritchett, C. J. & Grillmair, C. J. 1984 *Proc. astronom. Soc. Pacific* **96**, 349.

Ridgway, S. T., Carbon, D. F. & Hall, D. N. B. 1978 *Astrophys. J.* **225**, 138.

Salama, F. & Allamandola, L. J. 1992 *Nature, Lond.* **358**, 42.

Sarre, P. J. 1991 *Nature, Lond.* **351**, 356.

Scarrott, S. M., Watkin, S., Miles, J. R. & Sarre, P. J. 1992 *Mon. Not. R. astronom. Soc.* **255**, 11P.

Schmidt, G. D., Cohen, M. & Margon, B. 1980 *Astrophys. J.* **239**, L133.

Schmidt, G. D. & Witt, A. N. 1991 *Astrophys. J.* **383**, 698.

Sellgren, K. 1990 In *Dusty objects in the universe* (ed. E. Bussoletti & A. A. Vittone), pp. 35. Dordrecht: Kluwer.

Sellgren, K., Tokunaga, A. T. & Nakada, Y. 1990 *Astrophys. J.* **349**, 120.

Spitzer, L. 1978 In *Physical processes in the interstellar medium.* New York: J. Wiley.

Tamura, M., Hasegawa, T., Ukita, N., Gatley, I., McLean, I. S., Burton, M. G., Rayner, J. T. & McCaughrean, M. J. 1988 *Astrophys. J.* **326**, L17.

Tokunaga, A. T., Nagata, T., Sellgren, K., Smith, R. G., Onaka, T., Nakada, Y., Sakata, A. & Wada, S. 1988 *Astrophys. J.* **328**, 709.

Treffers, R. & Cohen, M. 1974 *Astrophys. J.* **188**, 545.

Truong-Bach, Nguyen-Q-Rieu, Omont, A., Olofsson, H. & Johansson, L. E. B. 1987 *Astronom. Astrophys.* **176**, 285.

Waelkens, C., Van Winckel, H., Trams, N. R. & Waters, L. B. F. M. 1992 *Astronom. Astrophys.* **256**, L15.

Wallerstein, G. 1973 *A. Rev. Astronom. Astrophys.* **11**, 115.

Wdowiak, T. J., Donn, B., Nuth, J. A., Chappelle, E. & Moore, M. 1989 *Astrophys. J.* **336**, 838.

Weidemann, V. & Koester, D. 1983 *Astronom. Astrophys.* **121**, 77.

Witteborn, F. C., Strecker, D. W., Erickson, E. F., Smith, S. M., Goebel, J. H. & Taylor, B. J. 1980 *Astrophys. J.* **238**, 577.

Woodward, C. E., Pipher, J. L., Shure, M., Forrest, W. J. & Sellgren, K. 1989 *Astrophys. J.* **342**, 860.

Zuckerman, B. & Aller, L. H. 1986 *Astrophys. J.* **301**, 772.

Zuckerman, B. & Dyck, H. M. 1980 *Astronom. Astrophys.* **209**, 119.

Discussion

A. S. Webster (*Royal Observatory, Edinburgh, U.K.*). There is now one carbon star known, and in the outflow from that star, the diffuse interstellar bands are known. So there is a binary companion between the spectra.

M. Jura. A problem arises, namely that the star lies right in the galactic plane, is eight-tenths of a degree from the plane, and the separation between the companion and the primary is little more than an arc-second, and the mass loss rate from that carbon star is not especially large (*ca.* $2 \times 10^{-7} M_{\odot}\ a^{-1}$). When you work out the amount of material nominally for standard, of course the interstellar medium (ISM) fluctuates: there is *ca.* 100 times as much material between us and that secondary, as circumstellar matter. So I am not convinced. But there is so much more interstellar compared with circumstellar matter, that unless the circumstellar matter is good at carrying diffuse bands, which it might be, it is probably more of an anomaly in the local ISM.

A. S. Webster. It may or may not be, but the spectrum is peculiar. It doesn't look like a normal interstellar spectrum.

M. Jura. Even in the ISM there are fluctuations in the strengths of the interstellar bands with respect to each other. When you look at different stars, they are families of interstellar bands, that don't all arise the same way. Even in the ISM, I think there are families that come together, but the relative amount of each family varies from one line of site to another in the ISM. So that may be what is going on in this particular object. That is my suspicion, but I can't prove it one way or the other.

R. C. Haddon (*AT&T Bell Laboratories, U.S.A.*). What are the prospects for the killer experiment, which will once and for all answer this long standing question on C_{60} in space.

M. Jura. C_{60} in space would be identified if, for example, we could see two, three or four of the IR bands that Krätschmer showed earlier, or if we could see either an absorption or an emission in some direction. We have tried without success, but we'll keep trying, because we are getting smarter as we think about it, and why we have failed in the past. That would be a definite way. It could be that much of the C_{60} is ionized, and then the diffuse bands would be slightly shifted in wavelength. Another feature which I would hope to learn about from somebody, is what happens when C_{60} absorbs a UV photon, and is free, so that it's not colliding with the walls of the experiment? How is that energy going to be re-radiated? Maybe someone here knows the answer or is able to figure it out, and then the astronomers might be able to tell you. For example, in the IR spectrum of IRC+10216, you can look at the wavelengths at 7 μm, 8.5 μm and so on, and you don't see an absorption; but these are low *f*-value transitions, so the absence of C_{60} is not really definitive. If you can figure this out, may an electronic transition or transitions, can be measured or calculated.

E. Wasserman (*The Du Pont Company, U.S.A.*). In the case of the polyaromatic hydrocarbons, do you think they come from the larger small molecules, or more from monoatomic, diatomic and triatomic sources?

M. Jura. In some sense that must be true, because it all comes out from a stellar atmosphere, where the temperatures are so high that it is, at most, diatomic. So originally the stuff is very simple, when it gets further out, say a few stellar radii, you might get grains, and I don't know then whether you are knocking things apart to get PAHs, or where you never actually build it up sufficiently large to get a drain. I don't know the answer to that.

Elemental carbon as interstellar dust

By C. T. Pillinger
Planetary Sciences Unit, Department of Earth Sciences, Open University, Milton Keynes MK7 6AA, U.K.

C_{60} has not yet been detected in primitive meteorites, a finding that could demonstrate its existence in the early solar nebular or as a component of presolar dust. However, other allotropes of carbon, diamond and graphite, have been isolated from numerous chondritic samples. Studies of the isotopic composition and trace element content and these forms of carbon suggest that they condensed in circumstellar environments. Diamond may also have been produced in the early solar nebula and meteorite parent bodies by both low-temperature–low-pressure processes and shock events. Evidence for the occurrence of another carbon allotrope, with sp hybridized bonding, commonly known as carbyne, is presented.

1. Introduction

At the same time that buckminsterfullerene was being conceived as a molecule of possible astrophysical significance, a number of much older forms of carbon were about to enjoy a new lease of life because of their discovery as presolar grains in primitive meteorites. Ever since the 1960s, it has been recognized that carbonaceous chondrites were a host for noble gases of anomalous isotopic composition (Anders 1981). The carriers of a litany of components, enjoying names such as Xe(HL) (also called CCF-Xe), s-Xe, Ne-E(L), Ne-E(H), etc., were believed to be unidentified carbon species called C_δ, C_β, C_α and C_ϵ respectively, themselves exhibiting unusual or exotic isotopic compositions (Swart *et al.* 1983*a*; Carr *et al.* 1983). In 1987, C_δ was shown to be diamond (Lewis *et al.* 1987) the meteorite mineral which contained Xe(HL) and nitrogen whose isotopic composition was greatly enriched in the light isotope ^{14}N (Lewis *et al.* 1983). This was almost immediately followed by the recognition (Bernatowicz *et al.* 1987) of C_β and C_ϵ as SiC, with over a factor of two ^{13}C more than the average solar system concentration. A little later came the identification of graphite, also ^{13}C enriched, as the Ne-E(L) related phase C_α (Amari *et al.* 1990). Other minor carbon containing entities, with non-descriptive but operationally defined names, e.g. C_λ (Ash *et al.* 1988), $C_\aleph$, C_κ (Ash *et al.* 1990), C_θ (Tang *et al.* 1987), H-C_δ (Verchovsky *et al.* 1992) have been suggested as presolar from their isotopic compositions or their associations but as yet absolute identifications are lacking. As SiC is not a carbon allotrope, a discussion of its importance as a presolar grain is not relevant in the context of this meeting and is given elsewhere (Pillinger & Russell 1993). Herein, the elemental carbon components which have been encountered in primitive meteorites or which have a related provenance are considered.

Phil. Trans. R. Soc. Lond. A (1993) **343**, 73–86
Printed in Great Britain

2. Separation and analysis of carbon components from meteorites

Elemental forms of carbon make up only a small part of the fabric of meteorites. They are hidden within a morass of organic matter, much of it indigenous, but some contamination added to the extraterrestrial samples accidentally during their sojourn on Earth. Fortunately the species of interest are robust and can withstand very harsh processing to isolate them from their immediate habitat. Thanks to the pioneering work of Tang *et al.* (1988), all the forms of carbon we are interested in can be extracted from meteorites with mineral acids such as HF/HCl and oxidizing conditions involving reagents like H_2O_2, NaOCl and $Cr_2O_7^{2-}$. To concentrate diamond $HClO_4$ is also used; only diamond is believed to be stable in $HClO_4$, but this is not an immutable fact. Much of the work discussed in this paper has used a technique known as stepped combustion (Swart *et al.* 1983*b*). Therein samples are heated through a series of temperatures, in the presence of oxygen, so that they are converted to carbon dioxide and released in order of increasing stability. Stepped combustion has been applied extensively to residues prepared up to and including $HClO_4$ stage of treatment. A derivative of stepped combustion, preparative precombustion, has been used with samples treated with HF/HCl (Ash *et al.* 1990). The carbon converted to CO_2 during any combustion reaction can be analysed for its isotopic composition using mass spectrometry. Specialized techniques have had to be developed to measure the smallest samples used (Carr *et al.* 1986; Prosser *et al.* 1990; Yates *et al.* 1992*a*, *b*). Carbon phases almost always co-exist with nitrogen which is released simultaneously during the combustion; because N exists in only trace abundance, static mass spectrometers (Wright *et al.* 1988) and appropriate handling techniques (Boyd *et al.* 1988) have had to be developed specifically for the investigations discussed here. Below, some isotopic data acquired by ion probe analysis are also included. However, this technique has a restricted application except in cases where individual grains can be located and mounted for study.

3. The occurrence of diamond in meteorites

Diamonds were first recognized in extraterrestrial materials when, within the space of a year or two, separate researchers reported their presence in the Novo Urei ureilite (Ierofeieff & Latchinoff 1888) and two iron meteorites, Magura (Weinschilling 1889) and Canyon Diablo (Foote 1891). During the next hundred years other ureilites (Vdovykin 1970) and at least one more iron meteorite (Clarke *et al.* 1981) were found to contain diamond. It was generally considered that diamond (or maybe more correctly, lonsdaleite, the hexagonal crystalline form of mineral) was a transformation product of graphite induced by impacts on Earth or in space (Lipschutz 1964). The diamonds discussed below have all been found in unshocked primitive meteorites, where the possibility of high-temperature–high-pressure reactions are precluded.

(a) C_δ *nanometre-sized, interstellar diamonds*

The diamonds first recognized by Lewis *et al.* (1987) have now been traced to every class of chrondritic meteorite (Alexander *et al.* 1990; Huss 1990; Russell *et al.* 1991) and are known to occur as a component within the matrix. The amounts available for extraction reach a maximum of 900 ppm but decline to nothing in higher petrologic grades. The cut-off point in unequilibrated ordinary chondrites is about type 3.6 but for enstatite chondrites the type 4s still contain diamond. Progressive

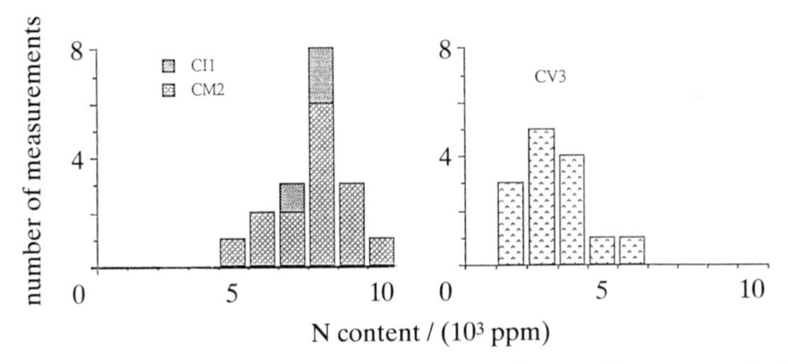

Figure 1. Nitrogen abundance in diamonds from various meteorite types.

metamorphism is believed to be the process responsible for destroying diamonds on the meteorite parent bodies; some possible metamorphic reactions have been proposed (Alexander *et al.* 1990).

A feature of the diamonds found in primitive meteorites, indeed the property which led to their isolation, is their very small particle size and ability to form a colloidal suspension in alkaline solutions. The particles precipitated from aqueous ammonia exhibit crystallites which have an average size of less than 3 nm (Lewis *et al.* 1989) equivalent to only *ca.* 2000 atoms of carbon. Electron diffraction patterns (Fraundorf *et al.* 1989; Gilkes *et al.* 1992) confirm the diamonds are cubic rather than hexagonal; the latter would be expected if shock was involved in their production. The particle size distribution is log-normal which suggests that either the grains have been size sorted or were in a state of growth rather than comminution (Lewis *et al.* 1989). It has been proposed that the apparent acidity of nanodiamonds is a property conveyed by carboxylic groups situated at corner sites on a cubic structure. Such a model is in surprisingly good agreement with data obtained by differential titration (Lewis *et al.* 1989). A question that has to be asked is, are the carboxyl groups original or a happenstance caused by harsh oxidative acid treatments used in isolation of the diamond? An attempt made to oxidize terrestrial microdiamonds to convert any peripheral dangling C atoms to carboxyl groups proved unsuccessful (Russell *et al.* 1992*a*). Of course, it is not possible to duplicate the conditions under which nanodiamond might occur, for example, as a continuous intergrowth with less robust forms of carbon which could be degraded to acid groups.

Among the most novel properties of C_δ is its ability to incorporate nitrogen. Terrestrial diamonds mined from kimberlite/lamproite source rocks, or found as placers, on occasion have nitrogen contents as high as 3000 ppm (Boyd *et al.* 1987) but the majority of samples contain 1000 ppm or much less (Deines *et al.* 1987). The diamonds from primitive chondrites have N abundances which extend from 2000–10000 ppm (figure 1) (Russell *et al.* 1991). Although there is a meteorite type dependent pattern in the data, models that attempt to explain that the results on the basis of gas loss, during simple metamorphic processing, do not work (Russell *et al.* 1992*b*). The major problem, and a quite remarkable feature of the N concentrations within the diamond, is the uniformity of N isotopic composition, the average of 22 determinations being $-343 \pm 16\%$ (Russell *et al.* 1991). (Herein isotopic compositions are discussed using the δ convention as enrichments or depletions relative to a standard, where for example $\delta^{13}\,C(‰) = 1000\,[(^{13}C/^{12}C)_{sample} - (^{13}C/^{12}C)_{std}]/(^{13}C/^{12}C)_{std}$.) If metamorphism is the cause of diminishing N content, it must be that high N diamonds are preferentially destroyed (Fisenko *et al.* 1992) without major isotopic fractionation. Whenever diamond-rich fractions are step-combusted the earliest temperature fractions have higher C/N ratios suggesting that either the finest grains or the surfaces of larger particles are nitrogen depleted (Russell *et al.* 1991) or perhaps have another carbon containing component present. At the same time the carbon isotopic compositions tend towards values enriched in the ^{12}C and nitrogen toward values enriched in ^{14}N. The shift in carbon isotopic composition is no more than 5‰ whereas $\delta^{15}N$ changes by more than 300‰. Such observations are indicative that a 'pure' phase is not being sampled thus C_δ may be only a hiding place for other minor but more interesting components. It is, however, difficult to envisage what the co-existing material is unless it is another form of elemental carbon resistant to the reagents used. It is claimed that diamond-like carbon is observed in the electron micrographs (Fraundorf *et al.* 1989; Blake *et al.* 1988) but a test carried out on a-C synthesized by a chemical vapour deposition (CVD) process revealed that such material is easily destroyed by $HClO_4$ (Pillinger *et al.* 1989); the apparently amorphous material seen may be fine-grained crystalline diamond incorrectly oriented for lattice imaging (Gilkes *et al.* 1992). As several investigators have suggested that presolar diamond might form around stars by a process akin to CVD (Lewis *et al.* 1987; Pillinger *et al.* 1989) an investigation of various samples made by this method has been performed. Fine-grained, micrometre-sized, CVD diamonds produced in a methane/hydrogen plasma burn sharply at relatively low temperatures. A sample made by sputtering a graphite target with argon gave residue, after treating with $Cr_2O_7^{2-}$ and $HClO_4$ to remove diamond-like material, which was enriched in ^{13}C by the order of 160‰ (Gilkes *et al.* 1992). An alternative suggestion (Blake *et al.* 1988) for the diamond formation mechanism is grain–grain collisions in the interstellar medium which transform amorphous carbon and graphite to the sp^3 hybrid allotrope. It may be relevant that diamonds have also been found in detonation soot (Blake *et al.* 1988; Greiner *et al.* 1988).

As already mentioned, C_δ is contaminated by Xe(HL) but the amounts of xenon are so small that only about one diamond crystal in 2×10^6 actually contains a xenon atom. The predominance of heavy and light xenon isotopes (hence the name Xe(HL)) requires an origin involving nucleosynthesis in a supernova (Clayton 1976); however, for diamonds to condense at all it is necessary to invoke their formation in a carbon star likely to be a red giant (Lewis *et al.* 1987). To accommodate these apparently incompatible conditions has required rather specific formation models to be invoked. One ingenious way to circumvent the problems has been to suggest origin in a binary star (Jorgensen 1988) another involves condensing diamonds in the He shell of a type II supernova, the only place in a massive star which is C-rich enough (Clayton 1989). Both the above proposals require that the xenon is incorporated into the diamond by a trapping mechanism probably ion implantation (Clayton 1981) after the crystal solidified and its trace content of nitrogen was fixed (Russell *et al.* 1991). No astrophysical means has yet been found to establish the diamond carbon isotopic composition at the observed $^{12}C/^{13}C = 93 \pm 0.5$ very close to the overall Solar System (probably 89); a carbon star origin would intuitively be ^{13}C enriched, the supernova hypothesis leads to ^{13}C content relative to ^{12}C of 10^{-4} (Clayton 1989). In some ways it would be much more satisfactory if C_δ could be further fractionated to separate Xe(H) and Xe(L), two components of different C isotope abundance and a specific nitrogen carrier; many have tried but so far to no avail (Schelhaas *et al.* 1990; Nichols *et al.* 1991; Lewis & Anders 1989; Russell *et al.* 1990).

Lewis *et al.* (1989) have discussed the possibility of detecting diamond in interstellar space. They argue this might only be achieved if C_δ like particles make up 10% of the total carbon budget. Allamandola *et al.* (1993) have observed a few percent of tertiary sp^3 carbon atoms in several protostars, possibly thereby confirming the predictions of Saslaw & Gaustad (1969) who were the first to suggest that diamond could be an interstellar grain.

(*b*) C_δ *in other places*

The fact that C_δ is such a resilient phase has encouraged the search to be carried to a variety of environments. For example, it has been known since the very earliest trace element analyses of the lunar regolith that a 1–2% (by mass) equivalent of carbonaceous chondrite debris survives from the great welter of impacts occurring over the last 3–4.5 billion years (Wasson & Baedecker 1970). To investigate whether any vestige of C_δ is included amongst this detritus, an HF–HCl residue of lunar soil was step combusted monitoring nitrogen isotopes as a fingerprint (Brilliant *et al.* 1992). Isotopically light nitrogen (−74‰) was observed over the correct temperature range to be indicative of about 2 ppm of relict interstellar grain material surviving. Unfortunately the normal method of concentrating diamond is inappropriate for experiments involving lunar samples because crystals 3 nm in size would be completely amorphized by continuous bombardment with the solar wind flux. Therefore, not surprisingly a sample of the HF/HCl residue processed with oxidizing acids afforded no evidence of residual diamond by TEM or isotopic fingerprinting. Experiments to prove the existence of interstellar diamond on the lunar surface will have to be more subtle.

Cosmic dust (micrometeorites less than 1 mm in size) are thought to be a likely source of primitive chondritic material. Such grains could be both asteroidal or cometary in origin. If the latter is true, then they might have brought to Earth copious quantities of presolar grains. Studying individual cosmic dust grains as though they were the equivalent of hand specimen conventional meteorites is a very exacting task but not without its rewards. In one instance a cosmic dust grain collected as part of the NASA high altitude programme was found to have lonsdaleite fragments (Rietmeijer & MacKinnon 1987). A much broader brush experiment has been attempted using samples of cryoconite (a sediment found in seasonal melt water pools on Greenland (Maurette *et al.* 1987)). After treatment of a specimen with HF/HCl, $Cr_2O_7^{2-}$ and $HClO_4$ according to the normal procedures for isolating diamond, the residue was step combusted using a special low blank procedure (Yates *et al.* 1992*a*, *b*). Although only 100 °C steps could be utilized, carbon dioxide liberated over the temperature régime 400–500° had an unusually low $\delta^{13}C$ of -34.4 ± 0.4‰, in keeping with the presence of C_δ. Transmission electron microscopy reveals the presence of a very fine grained diamond reminiscent of C_δ. Detailed isotope measurements are needed to resolve whether this has any relation to the material found in carbonaceous chondrites. In trying to estimate what the abundance of the diamond might be in cryoconite cosmic dust a number of assumptions have to be made. However, a concentration in excess of the maximum observed by Huss (1990) for CI matrix is arrived at suggesting unmelted micrometeorites might be very primitive.

Another environment that has been successfully searched for nanodiamonds is the Cretaceous–Tertiary boundary layer. Carlisle & Braman (1991) carried out the now time honoured acid dissolution procedure on samples from Knudsen's Farm, Alberta, Canada. They found 45 ppb of a white fraction 97% carbon almost entirely 3–5 nm diamonds. Without any corroborating evidence the authors jumped to the conclusion that the diamonds must have derived from a primitive meteorite in their estimation a C2 chondrite, thus claiming the first tangible evidence for a giant impact at the end of the Cretaceous. The presence of nanodiamonds has now been independently confirmed (Gilmour *et al.* 1991, 1992). However neither carbon isotopic composition or nitrogen abundance and $\delta^{15}N$ values are similar to values encountered for C_δ form of diamond in primitive meteorites. Thus Gilmour *et al.* (1992) favour an origin for the diamonds in the impact event or the associated fireball plasma. Despite these findings, Carlisle (1992) still adheres to an interpretation involving a direct meteorite input basing his arguments on an isotopic analysis for which no experimental details are available (Wright 1992).

(c) *Non-C_δ diamond*

Prospecting for nanodiamonds in meteorites has led to the discovery of two, maybe three other types of meteorite diamond. During an effort to explore the metamorphic stability of C_δ under reducing conditions the enstatite (EH4) chondrite Abee was studied (Russell *et al.* 1992*a*). It was anticipated that the relatively high grade meteorite would be virtually devoid of nanodiamonds; indeed it was, but amongst the oxidized acid residues, was found 100 ppm of diamond having a needle or lath-like morphology. The crystals that are 100 nm to 1 μm in length (i) burn at temperatures higher than C_δ (ii) can be extracted into alkaline solution, (iii) have $\delta^{13}C$ of -1.8 ± 0.24‰ and (iv) contain only 50 ppm N of $\delta^{15}N$ 0 to -20‰. The isotopic characteristics of Abee diamond are on the borderline of terrestrial kimberlite/lamproite stones (Galimov 1991; Boyd *et al.* 1987). Since the meteorite sample studied had surfaces which had been cut in undocumented circumstances, there was some concern about contamination by fragments from the saw blade, although every effort to exclude such a possibility was made. The evidence, however, that totally eliminates a terrestrial origin comes from noble gas measurements. Abee diamonds contain abundant cosmogenic neon which not only suggests they are indigenous to the meteorite but argues that the crystals were well distributed during cosmic irradiation. None of the information available for the new diamonds indicates a presolar origin, and since Abee is also unshocked it may transpire that low temperature low-pressure synthetic processes were active in the early solar nebula. Interestingly the morphology of Abee diamond resembles needle like crystals made by CVD (Seitz 1992). Abee diamonds could just be related to those which have been known about in ureilites for almost 100 years (Russell *et al.* 1992*a*).

Another meteorite which has been shown to be a new source of diamond is the very unusual chondrite Acfer 182 found in the Sahara. This meteorite is a reasonably large sample of class of meteorite first encountered as an 11 g pebble on Antarctica; it is distinguished by abundant small chondrules, calcium aluminium rich inclusions, 15% (by vol.) metal and the highest $\delta^{15}N$ values encountered for a bulk meteorite (Grady & Pillinger 1990, 1992, 1993). As Acfer 182 is a primitive chondrite it was surveyed for the presence of C_δ which is indeed present (Grady & Pillinger 1993). However, in the residue from the usual acid treatments, larger diamonds have also been found including a crystal almost 10 μm across with distinctive radial growth features suggesting a shock transformation. Further separations will be needed in order to obtain isotopic data for this new group of diamonds.

Stepped combustion analysis of 'pure' C_δ diamond fractions separated as colloids in ammonia are always accompanied by a small amount of carbon characterized by a slightly elevated $\delta^{13}C$. This additional material burns at temperatures higher than the main portion of the sample (Ash *et al.* 1987). Recently, it was found that diamond-rich residues particularly for the meteorites, Efremovka, Allende, Inman and Krymka contain a xenon component which is released by combustion at the same temperature as this carbon (Verchovsky *et al.* 1991, 1992). The isotopic composition of the new xenon, which has been named H-C_δXe (H denoting high temperature), is quite distinct from all other forms of the noble gas seen earlier (Verchovsky 1992). Gas of this isotopic composition had however been spotted (Lewis & Anders 1989) during one of the numerous attempts to separate Xe(H) and Xe(L) but at that time the possibility of another carbon component of higher stability than C_δ but associated with it was not considered. It is now very clear that the more stable carbon is 1000 ppm of the C_δ residue and its Xe (and other gases) content is much higher than the nanodiamonds (Verchovsky *et al.* 1992). The great stability of H-C_δ, i.e. resistance to the most oxidising acids suggests that it may be some other form of diamond. An interesting observation in this respect is that larger diamond crystallites have occasionally been seen by TEM within C_δ or similar residues (Fraundorf *et al.* 1989; Gilkes *et al.* 1992*a*). Such an observation is consistent with C_δ affording diamond X-ray diffraction data (H. J. Milledge, personal communication).

3. Other elemental forms of carbon in primitive meteorites

One of the enormous benefits of seeking diamond, or for that matter SiC, as presolar grain material is the incomparable resistance of the mineral to chemical attack. Other species are far more difficult but not impossible to isolate.

(*a*) *Graphite*

From the very earliest stepped combustion experiments performed on meteorite acid residues, it was known (Carr *et al.* 1983) that the carrier of Ne-E(L) burnt at a temperature somewhere between that of Xe(HL) and s-Xe/Ne-E(H). A sample enriched in Ne-E(L) liberated a small amount of carbon with $\delta^{13}C = +340‰$ which burned at around 650 °C. Very similar $\delta^{13}C$ values were encountered at about this temperature during stepped combustions of a variety of samples, processed in different ways, surprisingly including some treated with $HClO_4$. The evidence seemed to point strongly to the conclusion that C_α, the Ne-E(L) carrier was only slightly enriched in ^{13}C. The same carbon isotopic composition was also observed at approximately 650 °C in an HF/HCl residue of the Allende meteorite which had been extensively precombusted to remove low stability amorphous carbon (Ash *et al.* 1990). A problem with equating this result with the presence of C_α is that Allende is a meteorite which contains little if any Ne-E(L) (Frick & Pepin 1981). To reconcile the data Ash *et al.* (1990) suggested that C_α might be degassed in Allende; a new name $C_\aleph$ was introduced to distinguish the gas-poor from the gas-rich component. Interestingly, when carried to extreme the precombustion experiment revealed for the first time the existence of carbon enriched in ^{12}C (Ash *et al.* 1988). A minute component (only 0.1 ppm) burning at about the correct temperature for graphite combustion became apparent after many hours of precombustion.

In an effort to track the carrier of Ne-E(L) further, Amari *et al.* (1990) dispensed with the use of the most powerful oxidants (i.e. $HClO_4$) substituting NaOCl and alkaline H_2O_2 to remove stubborn organic phases. After treatment a residue from the Murchison meteorite was separated according to density. Ne-E(L) was found specifically in a sub-class of dense, rounded graphite grains 1.5–6 μm in size, the most convincing experiments being carried out using a focused laser to degas individual grains for mass spectrometric studies (Nichols *et al.* 1992). In addition to the dust grains, the residues contained other forms of graphite including aggregates of 0.1 μm spherules and even euhedral graphite crystals. Isotopic composition of various of these components have been measured by ion microprobe. Aggregates give normal solar system values for carbon but have abundant nitrogen up to 300‰ enriched in ^{15}N (Amari *et al.* 1990). The dense spheres are among the most exotic particles yet studied. Many have $^{12}C/^{13}C$ ratios which range from 0.09 to 16 times the solar system value of 89. The most ^{12}C enriched grains have a $^{12}C/^{13}C$ ratio of 1440, far in excess of values observed for carbon stars, which do not go beyond 97 (Lambert *et al.* 1986) although in principle these could be higher. It seems to be clear that the ubiquitous $\delta^{13}C$ of +340‰, measured for stepped combustion, is due to the averaging of all these various components. However, the mix seems to be amazingly consistent and it is difficult to understand why stepped combustion is not able to resolve at least partially the dense graphite spheres from the sub-micrometre sized aggregates.

(*b*) *Carbynes*

Like many other subjects the quest for the identification of presolar grains was not without its false starts. Material with polyyne or cummulene structures was first claimed to have been made in the 1960s, but in 1968, El Goresy and Donnay found 'white carbon' with a distinctive X-ray diffraction pattern within a graphite-rich gneiss from the Ries Meteorite crater in Bavaria. This new allotrope of carbon was given the mineral name chaoite. Shortly after it was identified in a meteorite, the Havero ureilite (Vdovykin 1972). During the late 1970s the number of possible carbynes grew as species of different chain lengths were proposed (Whittaker & Wolten 1972). About this time also cyano polyacetylenes of increasing chain length were being identified as interstellar molecules (Morris *et al.* 1976). With various pieces of circumstantial evidence accumulating (Whittaker *et al.* 1981) proposed carbynes as anomalous noble gas host phases; Webster (1981) suggested they might contribute to interstellar dust. The idea that the C_δ, C_β, etc., might be almost entirely carbynes was very soon refuted by detailed TEM studies (Smith & Buseck 1981*a*); indeed the notion that carbynes were the third crystalline allotrope of carbon was questioned (Smith & Buseck 1982).

Although the possibility of carbynes in primitive meteorites fell into disrepute for ten years, a considerable literature on the existence of an allotrope involving sp hybridized carbon, mostly Russian, but also Western (Heimann *et al.* 1983, 1984) has built up in the subject. Kudryavtsev *et al.* (1992) give single crystal X-ray data. While attempting to study C_δ with high resolution TEM (Gilkes *et al.* 1992) found crystalline regions with interplanar spacings higher than diamond, silicon carbide and graphite. Selected area electron diffraction studies and lattice images from several of these particles match figures obtained for chaoite. EDX spectra confirm that silicon is absent so that the grains are not the contaminants suspected in earlier studies (Smith & Buseck 1981*a*, *b*, *c*, 1982). Repeated attempts have been made to obtain electron energy loss spectra (EELS) for specific particles fully identify some kind of carbyne. These experiments have been frustrated because of the instability of the tiny grains in the electron beam. This, however, might itself be indicative that the particles are the sp allotrope by analogy to the work of Kudryavtsev (1992). EELS data with a characteristic K edge distinct from amorphous carbon, graphite and diamond have been obtained from unspecified areas; the absence of Si edges again rules out the possibility of contamination by silicates. Similar results TEM and EELS data have been obtained from putative carbyne material deposited in the laboratory during a plasma arc experiments (Gilkes & Pillinger, unpublished results).

(c) C_{60} *in Meteorites*

Almost as soon as buckminsterfullerene was proposed as a form of carbon, the idea of searching for the C_{60} structure as an interstellar grain in meteorites was suggested (Heymann 1986). Indeed C_{δ} nanometre-sized diamonds were found as a result of an erroneous 'hunch' that C_{60} might be the carrier of Xe(HL) (Lewis *et al.* 1987). The isolation of buckminsterfullerene in quantity has allowed its properties to be evaluated (Gilmour *et al.* 1991; Heymann 1991) so that a more directed search may be conducted to investigate its role as a presolar grain constituent of primitive meteorites. Tests performed with authentic C_{60} samples show that like diamond and SiC, buckminsterfullerene is stable in all the reagents used to isolate interstellar grains from meteorites including oxidising acids, even $HClO_4$. With respect to combustion properties C_{60} burns at variable temperatures but this is probably a feature of the crystal size rather than molecular dimensions. It seems likely that C_{60} would burn at or around the same temperature as nanometre sized diamonds; the lowest combustion temperatures observed with authentic samples being *ca.* 400 °C. Interestingly, isotopic compositions measured during stepped combustions indicate that the $\delta^{13}C$ for C_{60} is about 12‰ heavier than the likely isotopic composition of the graphite rods from which it was made. This may reflect an isotopic fractionation during the synthesis or could be due to the purification during HPLC extraction.

Although C_{60} itself has not yet been discovered, graphite particles similar to those reported by Ugarte (1992) have been found in HF/HCl by insoluble residues of meteorites such as Allende (Smith & Buseck 1981*b*) and Adrar 003 (Gilkes & Pillinger, unpublished results); whether these are seeded by C_{60} is an open question.

Although it is necessary to continue to look for C_{60} in primitive meteorites as an interstellar grain, it might be more appropriate to seek its hydrogenated counterparts since it is difficult to find astrophysical environments which are hydrogen free. Nothing is yet known about the properties of species of the form $C_{60}H_n$ under the conditions prevailing during extraction of primitive meteorites. The solubility of C_{60} in organic solvents means that in the past it might have been removed from acid residues during the work up treatment, particularly in samples extracted with CS_2 to remove sulphur. Acid residues of various meteorite have been extracted with dichloromethane and extracts analysed by HPLC. None of these samples showed any evidence of C_{60}. Nevertheless infrared adsorption bands attributable to C_{60} have been observed in the interstellar spectrum (see other contributions in this volume) so the search will continue. Interestingly C_{60} has been identified in a terrestrial Precambrian coal from the Karelia region of Russia. Such a material will represent a good model for meteorite studies.

4. Conclusion

The first steps in any new discipline are always slow and tentative but once they have been made, growth of the subject is almost always exponential. Such has been the case for C_{60} and the discovery of interstellar grains in meteorites. Both these subjects are new areas of science involving carbon, demonstrating the spectacular scope in the chemistry of this unique element.

References

Alexander, C. M. O., Arden, J. W., Ash, R. D. & Pillinger, C. T. 1990 Presolar components in ordinary chondrites. *Earth Planet. Sci. Lett.* **99**, 220–229.

Allamandola, L. J., Sandford, S. A., Tielens, A. G. G. M. & Hirbst, T. M. 1993 Infrared spectroscopy of dense clouds in the C-H stretch region: Methanol and diamonds. *Astrophys. J.* (Submitted.)

Amari, S., Anders, E., Virag, A. & Zinner, E. 1990 Interstellar graphite in meteorites. *Nature, Lond.* **345**, 238–240.

Anders, E. 1981 Noble gases in meteorites: Evidence for presolar matter and heavy elements. *Proc. R. Soc. Lond.* A **374**, 207–238.

Ash, R. D., Grady, M. M., Wright, I. P., Pillinger, C. T., Anders, E. & Lewis, R. S. 1987 An investigation of carbon and nitrogen isotopes in C_δ and the effects of grain size upon combustion temperature. *Meteoritics* **22**, 319.

Ash, R. D., Arden, J. W., Grady, M. M., Wright, I. P. & Pillinger, C. T. 1988 An interstellar dust component rich in ^{12}C. *Nature, Lond.* **336**, 228–230.

Ash, R. D., Arden, J. W., Grady, M. M., Wright, I. P. & Pillinger, C. T. 1990 Recondite interstellar carbon revealed by preparative precombustion. *Geochim. Cosmochim. Acta* **54**, 455–468.

Bernatowicz, T., Fraundorf, G., Tang, M., Anders, E., Wopenka, B., Zinner, E. & Fraundorf, P. 1987 Evidence for interstellar SiC in the Murray carbonaceous meteorite. *Nature, Lond.* **330**, 728–730.

Boyd, S. R., Mattey, D. P., Pillinger, C. T., Milledge, H. J., Mendelssohn, M. J. & Seal, M. 1987 *Earth Planet. Sci. Lett.* **86**, 341–353.

Boyd, S. R., Wright, I. P., Franchi, I. A. & Pillinger, C. T. 1988 Preparation of sub-nanomole quantities of nitrogen for stable isotope analysis. *J. Phys.* E **21**, 866–885.

Blake, D. F. *et al.* 1988 The nature and origin of interstellar diamond. *Nature, Lond.* **332**, 611–613.

Brilliant, D. R., Franchi, I. A., Arden, J. W. & Pillinger, C. T. 1992 An interstellar component in the lunar regolith. *Meteoritics* **27**, 206–207.

Carlisle, D. R. 1992 *Nature, Lond.* **357**, 119.

Carlisle, D. B. & Braman, D. R. 1991 Nanometre-sized diamonds in the Cretaceous/Tertiary boundary clay of Alberta. *Nature, Lond.* **352**, 708–709.

Carr, R. H., Wright, I. P., Joines, A. W. & Pillinger, C. T. 1986 Measurement of carbon stable isotopes at the nanomole level: a static mass spectrometer and sample preparation techniques. *J. Phys.* E **19**, 798–808.

Carr, R. H., Wright, I. P., Pillinger, C. T., Lewis, R. H. & Anders, E. 1983 Interstellar carbon in meteorites: Isotopic analysis using static mass spectrometry. *Meteoritics* **18**, 277.

Clarke, R. S., Appleman, D. E. & Ross, D. R. 1981 An Antarctic meteorite contains preterrestrial impact-produced diamond and lonsdaleite. *Nature, Lond.* **214**, 396–398.

Clayton, D. D. 1976 Spectrum of carbonaceous chondrite fission xenon. *Geochim. Cosmochim. Acta* **40**, 563–565.

Clayton, D. D. 1981 Some key issues in isotopic anomalies: Astrophysical history and aggregation. *Proc. Lunar Planet. Sci. Conf.* B **12**, 1781–1802.

Clayton, D. D. 1989 Origin of heavy xenon in meteoritic diamonds. *Astrophys. J.* **340**, 613–619.

Deines, P., Harris, J. W., Spear, P. M. & Gurney, J. J. 1989 Nitrogen and ^{13}C content of Finsch and Premier diamonds and their implications. *Geochim. Cosmochim. Acta* **53**, 1367–1378.

El Goresy, A. & Donnay, G. 1968 A new allotropic form of carbon from the Ries crater. *Science, Wash.* **161**, 363–364.

Fisenko, A. V., Russell, S. S., Ash, R. D., Semjenova, L. F., Verchovsky, A. B. & Pillinger, C. T. 1992 *Lunar Planet. Sci.* **XXIII**, 365–366.

Foote, A. E. 1891 A new locality for meteorite iron and a preliminary notice of the discovery of diamonds in the iron. *Proc. Am. Ass. Adv. Sci.* **40**, 279–283.

Fraundorf, P., Fraundorf, G., Bernatowicz, T., Lewis, R. S. & Tang, M. 1989 Stardust in the T. E. M. *Ultramicrosc.* **27**, 401–412.

Frick, U. & Pepin, R. O. 1981 On the distribution of noble gases in Allende: a differential oxidation study. *Earth Planet. Sci. Lett.* **56**, 64–81.

Galimov, E. M. 1991 Isotopic fractionation related to kimberlite magmatism and diamond formation. *Geochim. Cosmochim. Acta* **55**, 1697–1708.

Ganapathy, R., Keays, R. R., Laul, J. C. & Anders, E. 1970 Trace elements in Apollo 11 lunar rocks: Implications for meteorite influx and origin of Moon. *Proc. Apollo Lunar Sci.* **2**, 1117–1442.

Gilkes, K. W. R., Gaskell, P. H., Russell, S. S., Arden, J. W. & Pillinger, C. T. 1992 Do carbynes exist in meteorites after all? *Meteoritics* **27**, 224.

Gilmour, I. *et al.* 1991 A search for the presence of C_{60} as an interstellar grain in meteorites. *Lunar Plant. Sci.* **XXII**, 445–446.

Gilmour, I., Russell, S. S. & Pillinger, C. T. 1992 Origin of diamonds in K/T boundary clays. *Lunar Planet. Sci.* **XXIII**, 1187–1188.

Gilmour, I., Russell, S. S., Arden, J. W., Lee, M. R., Franchi, I. A. & Pillinger, C. T. 1993 Terrestrial carbon and nitrogen isotope ratios from Cretaceous–Tertiary boundary micro-diamonds. *Science, Wash.* (In the press.)

Grady, M. M. & Pillinger, C. T. 1990 ALH 85085: nitrogen isotopic analysis of a highly unusual primitive chondrite. *EPSL* **97**, 29–40.

Grady, M. M. & Pillinger, C. T. 1992 The continuing search for the location of ^{15}N-enriched nitrogen in Acfer 182. *Meteoritics* **27**, 226–227.

Grady, M. M. & Pillinger, C. T. 1993 Acfer 182: Search for the location of ^{15}N enriched nitrogen in an unusual chondrite. *EPSL* (Submitted.)

Greiner N. R., Philips, D. S., Johnson, J. D. & Volk, F. 1988 Diamonds in detonation soot. *Nature, Lond.* **333**, 440–442.

Heymann, D. 1986 Buckminsterfullerene and its siblings and soot carriers of trapped inert gases in meteorites. Proc. 17th Lunar Planet Sci. Conf. *J. geophys. Res.* **19**, E135–138.

Heymann, D. 1991 The geochemistry of buckminsterfullerene (C_{60}) I. Solid solutions with sulfur and oxidation with perchloric acid. *Lunar Planet Sci.* **XXII**, 569–570.

Heimann, R. B., Kleiman, J. & Salansky, N. M. 1983 A unified structure approach to linear carbon polytypes. *Nature, Lond.* **306**, 164–167.

Heimann, R. B., Kleiman, J. & Salansky, N. M. 1984 Structural aspects and conformation of linear carbon polytypes (carbynes). *Carbon* **22**, 147–155.

Huss, G. R. 1990 Ubiquitous interstellar diamond and SiC in primitive chondrites: abundances reflect metamorphism. *Nature, Lond.* **347**, 159–162.

Ierofeieff, M. V. & Latchinoff, P. A. 1888 Météorite diamantére tombé le 10/22 Septembre 1886 en Russie, à Nowo Urei, governemet de Penze. *C. r. Hebd. Seanc. Acad. Sci. Paris* **106**, 1679–1682.

Jorgensen, V. G. 1988 Formation of Xe(HL) – enriched diamond grains in stellar environments. *Nature, Lond.* **332**, 702–705.

Kroto, H. W., Kirby, C., Walton, D. R. M., Avery, L. M., Broten, H. W., McLeod, J. M. & Oka, T. 1978 The detection of cyanohexatroyne $H(C{\equiv}C)_3CN$ in Heiles's Cloud 2. *Astrophys. J.* **219**, L133–137.

Kudryavtsev, Yu., Evsyukov, S. E., Babaev, V. G., Guseva, M. B., Khvostov, V. V. & Krechko, L. M. 1992 Oriented carbyne layers. *Carbon* **30**, 213–221.

Lambert, D. L., Gustafsson, B., Enksson, K. & Hinkle, K. H. 1986 *Astrophys. J. Suppl.* **62**, 373–425.

Lewis, R. S. & Anders, E. 1989 Xenon HL in diamonds from the Allende meteorite x-composite nature. *Lunar. Planet. Sci.* **XIX**, 679–680.

Lewis, R. S., Anders, E., Wright, I. P., Norris, S. J. & Pillinger, C. T. 1983 Isotopically anomalous nitrogen in primitive meteorites. *Nature, Lond.* **305**, 767–771.

Lewis, R. S., Tang, M., Wacker, J. F., Anders, E. & Steel, E. 1987 Interstellar diamonds in meteorites. *Nature, Lond.* **326**, 160–162.

Lewis, R. S., Anders, E. & Draine, B. T. 1989 Properties, detectability and origin of interstellar diamonds in meteorites. *Nature, Lond.* **339**, 117–121.

Lipschutz, M. E. 1964 *Science, Wash.* **143**, 1431–1433.

Maurette, M., Jéhanno, C., Robin, E. & Hammer, C. 1987 Characteristics and mass distribution of extraterrestrial dust from the Greenland Icecap. *Nature, Lond.* **328**, 699–702.

Morris, M., Turner, B. E., Palmer, P. & Zuckerman, B. 1976 Cyano acetylene in dense interstellar clouds. *Astrophys. J.* **205**, 82–93.

Nichols, R. H., Hohenberg, C. M., Alexander, C. M. O., Olinger, C. T. & Arden, J. W. 1991 Xenon and neon from acid resistant residues of Inman and Treschitz. *Geochim. Cosmochim. Acta* **55**, 2921–2936.

Nichols, R. H., Hohenberg, C. M., Hoppe, P., Amari, S. & Lewis, R. S. 1992 ^{22}Ne-E(H) and ^{4}He in single SiC grains and ^{22}Ne-E(L) in single C_α grains of known C-isotopic composition. *Lunar Planet. Sci.* **XXIII**, 989–990.

Pillinger, C. T. & Russell, S. S. 1993 Interstellar SiC grains in meteorites. *J. chem. Soc. Faraday Trans.* (In the press.)

Pillinger, C. T., Ash, R. D. & Arden, J. W. 1989 The rôle of CVD in the production of interstellar grains. *Meteoritics* **24**, 316.

Prosser, S. J., Wright, I. P. & Pillinger, C. T. 1990 A preliminary investigation into the isotopic measurement of carbon at the picomole level using static vacuum mass spectrometry. *Chem. Geol.* **83**, 71–88.

Rietmeijer, F. J. M. & MacKinnon, I. D. R. 1987 Metastable carbon in two chondritic porous interplanetary dust particles. *Nature, Lond.* **326**, 162–165.

Russell, S. S., Ash, R. D., Arden, J. W. & Pillinger, C. T. 1990 Nitrogen in diamond from primitive meteorites. *Meteoritics* **25**, 402.

Russell, S. S., Arden, J. W. & Pillinger, C. T. 1991 Evidence for multiple sources of diamonds. *Science, Wash.* **254**, 1188–1191.

Russell, S. S., Pillinger, C. T., Arden, J. W., Lee, M. R. & Ott, U. 1992*a* A new type of meteoritic diamond in the enstatite chondrite Abee. *Science, Wash.* **256**, 206–209.

Russell, S. S., Becker, R. H. & Pillinger, C. T. 1992*b* Modelling nitrogen degassing in chondrite diamonds. *Lunar Planet. Sci.* **XXIII**, 1187–1190.

Saslaw, W. C. & Gaustad, J. E. 1969 Interstellar dust and diamonds. *Nature, Lond.* **221**, 160–162.

Schelhaas, N., Ott, U. & Begemann F. 1990 Trapped noble gases in unequilibrated ordinary chondrites. *Geochim. Cosmochim. Acta* **54**, 2869–2882.

Seitz, R. 1992 *Sky Telescope* **84**, 247.

Smith, P. P. K. & Buseck, P. R. 1981*a* Carbon in the Allende meteorite: evidence for poorly graphitised carbon rather than carbyne. *Proc. Lunar. Sci. Conf.* **12**b, 1167–1175.

Smith, P. P. K. & Buseck, P. R. 1981*b* Graphitic carbon in the Allende meteorite: a microstructural study. *Science, Wash.* **212**, 322–324.

Smith, P. P. K. & Buseck, P. R. 1982 Carbynes forms of carbon: Do they exist? *Science, Wash.* **216**, 984–986.

Swart, P. K., Grady, M. M., Pillinger, C. T., Lewis, R. S. & Anders, E. 1983*a* Interstellar carbon in meteorites. *Science, Wash.* **220**, 406–410.

Swart, P. K., Grady, M. M. & Pillinger, C. T. 1983*b* A method for the identification and elimination of contamination during carbon isotopic analysis of extraterrestrial material. *Meteorites* **18**, 137–154.

Tang, M., Lewis, R. S., Anders, E., Grady, M. M., Wright, I. P. & Pillinger, C. T. 1988 Isotopic anomalies of Ne, Xe and C in meteorites. I. Separation of carriers by density and chemical resistance. *Geochim. Cosmochim Acta* **52**, 1221–1234.

Vdovykin, G. P. 1970 Ureilites. *Space Sci. Rev.* **10**, 483–510.

Vdovykin, G. P. 1972 Forms of carbon in the new Havero ureilite of Finland. *Meteoritics* **7**, 547–552.

Verchovsky, A. B., Ott, U., Russell, S. S. & Pillinger, C. T. 1991 A new Xe component in diamond rich acid residues from the Efremovka CV3 carbonaceous chondrite. *Meteoritics* **26**, 402.

Verchovsky, A. B., Russell, S. S., Pillinger, C. T., Fisenko, A. & Shukolyukov, Yu. A. 1992 Hunting for the carrier of a new Xe component (H-C_δXe) in diamond-rich residues of primitive meteorites. *Meteoritics* **27**, 301–302.

Wasson, J. T. & Baedecker, P. A. 1970 Ga, Ge, In, Ir and An in lunar, terrestrial and meteoritic basalts. *Proc. Apollo 11 Lunar Science Conf.* **2**, 1741–1750.

Weinschilling, S. J. 1889 Uber einige Bestandtaile des meteoreisens. *Von. Magura. Arva. Ungarn. Ann. des Naturhistorischen Hofmuseums, Wien* **4**, 93–100.

Whittaker, A. G. & Wolten, G. M. 1972 Carbon: A suggested new hexagonal crystal form. *Science, Wash.* **178**, 54–56.

Whittaker, A. G., Watts, E. J., Lewis, R. S. & Anders, E. 1981 Carbynes carriers of primordial noble gases in meteorites. *Science, Wash.* **269**, 1512–1514.

Wright, I. P. 1992 *Nature, Lond.* **358**, 198.

Wright, I. P., Boyd, S. R., Franchi, I. A. & Pillinger, C. T. 1988 Determination of high precision nitrogen stable isotope ratios at the sub-nanomole level. *J. Phys.* E **21**, 865–875.

Yates, P. D., Wright, I. P. & Pillinger, C. T. 1992*a* Application of high sensitivity carbon isotope techniques – a question of blanks. *Chem. Geol.* (*Isotope Geoscience*) **101**, 81–91.

Yates, P. D., Arden, J. W., Wright, I. P., Pillinger, C. T. & Hutchison, R. 1992*b* A search for presolar material within an acid resistant residue of Greenland cryoconite. *Meteoritics* **27**, 309–310.

Discussion

H. W. Kroto (*University of Sussex, U.K.*). A current controversy suggests that carbynes were due to the presence of an artefact. Furthermore, all the materials you have shown are essentially supposed to be condensed polyynes. I have yet to see any evidence to support the claim that they contain significant numbers of triple bonds.

C. T. Pillinger. We have endeavoured to prove that this might be carbyne by recording the electron energy loss spectrum. Unfortunately the material is very unstable in the electron beam required, so we have been unable to obtain any spectra from the regions where we believe such grains are. They may well differ from graphite, amorphous carbon and from other forms of carbon that we know about, including C_{60}; furthermore, they don't have silicon edges. So this is a carbonaceous phase, that has an electron diffraction pattern, which seems to match a material called chaoite, the 'white carbon' found in the Ries crater. I know you are worried about the stability of these samples. It is known that carbynes can be stabilized by attaching metals to the ends of the chains and one reason why I might now be able to have carbynes to hand is that extra atoms may be attached to the chains. This would possibly provide a link to the long chain triply-bonded carbon species containing nitrogen atoms. More information will be available if we obtain an EEL spectrum, which would tell us whether the material contained nitrogen or not. We are working on the problem, and I like to believe that about 20 years ago another allotrope of carbon was discovered.

R. C. Haddon (*AT & T Bell Laboratories, U.S.A.*). How much of this material do you actually possess? What about the surface of the diamond grains, presumably they have to be terminated with something?

C. T. Pillinger. We have enough material to see, about 30 grains. With a 200 atom grain, you must terminate the bonds on the surface. Now we know that hydrogen can terminate the bonds. I think there is an absorption at *ca.* 3 μm; the grains may therefore contain some hydrogen. The mean size for these diamond particles is 2000 atoms, although there are certainly diamonds which are smaller. Another point about the diamond grains is that they appear to have carboxyl groups attached, and this enables us to extract them from the meteorite. But whether the carboxyl groups are introduced as a result of our treatment, or whether they are there originally, is not clear. Characterizing nanocrystals is not very easy.

S. Iijima (*NEC Corporation, Japan*). Could you comment on other work in which C_{60} has been found in Nature?

C. T. Pillinger. It has been found in shungite; a Precambrian coal from Karelia. I have a sample of shungite which we have tried to extract with toluene to see whether we could obtain a spectrum with peaks corresponding to C_{60}. This is very recent data and our initial results are not promising.

S. Leach (*Observatoire de Paris-Meudon, France*). Do any of the diamonds contain colour centres due to metallic impurities?

C. T. Pillinger. There are no metal impurities in these diamonds high enough to give that kind of response.

The pattern of additions to fullerenes

By R. Taylor

School of Chemistry and Molecular Sciences, University of Sussex, Brighton BN1 9QJ, Sussex, U.K.

Conjugation in C_{60} is not as extensive as was originally anticipated because, for various reasons, the pentagon rings avoid containing double bonds. As a consequence, there is extensive bond localization and the molecule, which is quite reactive, and displays superalkene rather than superaromatic properties. C_{70} behaves in a similar fashion; other fullerenes may follow suit. Additions predominate and C_{60} is particularly susceptible to nucleophilic attack. Added groups may also be readily replaced by nucleophiles, although the reaction mechanism is uncertain at present. The functionalized molecule tends to revert to the parent fullerene at moderate temperatures, and characterization of reaction products by mass spectrometry is thus particularly difficult. This fact, coupled with the complexity of the addition products, makes work with fullerenes exacting. A selection of reactions studied to date and the progress made towards identifying various patterns of addition are described.

1. Introduction

When spectroscopic evidence for the existence of C_{60} was first obtained, a view prevailed that it would be a very unreactive molecule. This conclusion was based on the assumption that with a possible 12500 resonance structures (Klein *et al.* 1986), C_{60} would be superaromatic. However, the earlier molecular orbital calculations (Bochvar & G'alpern 1973) correctly predicted that there would be substantial bond fixation in the molecule; more recently, the bond lengths have been determined by neutron diffraction studies to be 1.391 Å and 1.455 Å (David *et al.* 1991); other methods give similar values.

Thus C_{60} is a superalkene rather than a superaromatic and readily undergoes additions. Bond fixation arises because structures with double bonds in pentagonal rings are unfavourable in carbocyclic chemistry, due probably to the increase in strain that would result from bond-shortening; this phenomenon is particularly well demonstrated by the bond lengths in corannulene (figure 1 (Barth & Lawton 1971)). The structure of C_{60} is such that avoidance of high-order bonds in the pentagonal rings results in complete bond alternation in each 'benzenoid' hexagonal ring. If fullerene isomers are drawn such that the double bonds are exocyclic to the pentagonal rings, then those which have the highest proportion of benzenoid rings tend to be the most stable ones (Taylor 1991, 1992*a*; Fowler 1992), however, this factor may be of diminished importance for higher fullerenes. A consequence of the favoured bond alternation is that the pentagonal rings ideally lie *meta* to one another (I, figure 2) as they each do in C_{60}. By contrast, if they lie either *ortho* or *para* to one another (II, III, figure 2), then at least one pentagonal ring must contain a high-order bond. It is noteworthy therefore that each of the numerous published mass

Phil. Trans. R. Soc. Lond. A (1993) **343**, 87–101

Printed in Great Britain

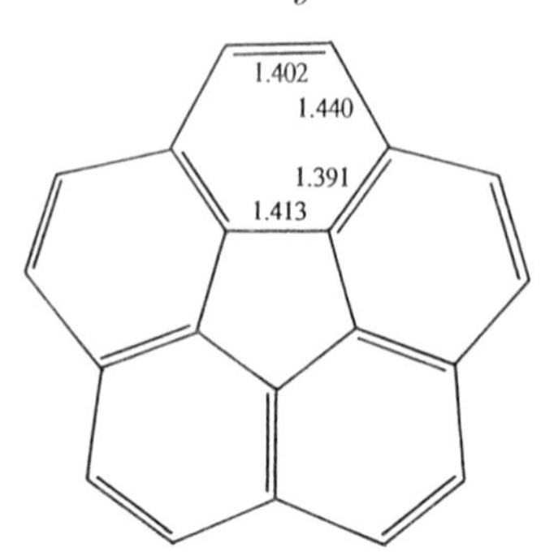

Figure 1. Bond lengths (in Å) in corannulene.

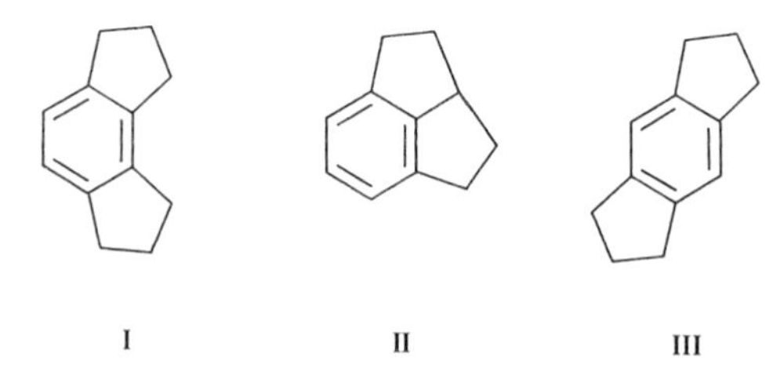

Figure 2. Dispositions of two pentagonal rings adjacent to a hexagonal one.

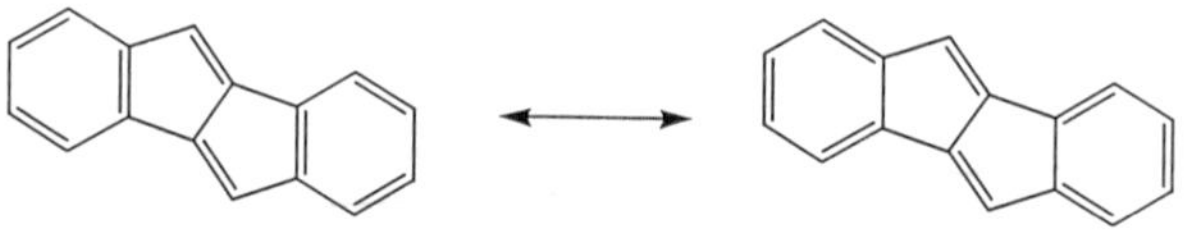

8π 16π

Figure 3. Antiaromaticity introduced by adjacency of pentagonal rings.

spectra of higher fullerenes show C_{80} to be particularly unstable; in the icosahedral isomer, all pentagonal rings lie in the *para* relation (Taylor 1992*a*). The unfavourability of the *ortho* arrangement also provides an explanation of the isolated pentagon rule (IPR), i.e. the most stable fullerenes are those with isolated pentagons (Schmalz *et al.* 1986, 1988), interpreted originally in terms of minimization of strain (Kroto 1987). The presence of adjacent pentagons is unfavourable because this leads to regions of antiaromaticity within the molecules (figure 3 (Schmalz *et al.* 1986, 1988; Taylor 1991).

The sp^2-hybridized carbon atoms of the fullerene cages are more electronegative than sp^3-hybridized carbon and have electron-withdrawing ($-I$) effects (Bent 1961). This creates a tendency for fullerenes to react with nucleophiles, an additional reason for this being that addition of a nucleophile converts one of the pentagonal rings of the pyracyclene units within the molecule (figure 4) into a 6π-aromatic system. The high electron affinity and consequent oxidizing behaviour of C_{60} likewise results from the addition of two electrons to give two aromatic pentagonal rings (figure 4) (Wudl 1992). The presence of six pyracyclene units in C_{60} accounts for the observed formation of polyanions (Bausch *et al.* 1991).

The electron-deficient nature of C_{60}, and its behaviour as an alkene rather than as an aromatic was clearly shown by reaction with η^5-$C_5Me_5Ru(MeCN)_3^+(CF_3SO_3^-)_3$, a reagent which invariably reacts with aromatic rings, losing all three cyanomethane

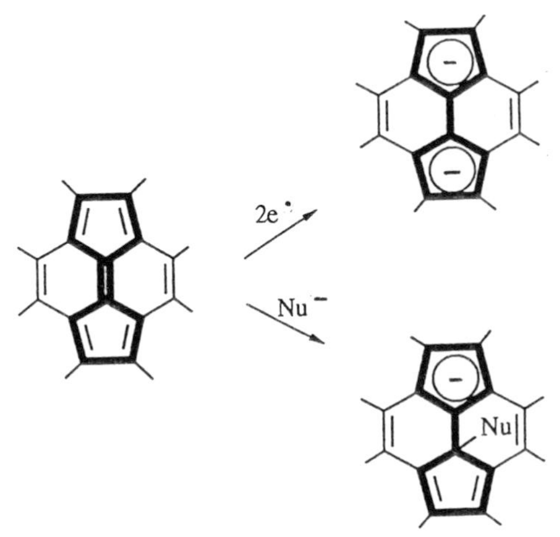

Figure 4. Addition of either two electrons, or a nucleophile to a pyracyclene unit of C_{60}.

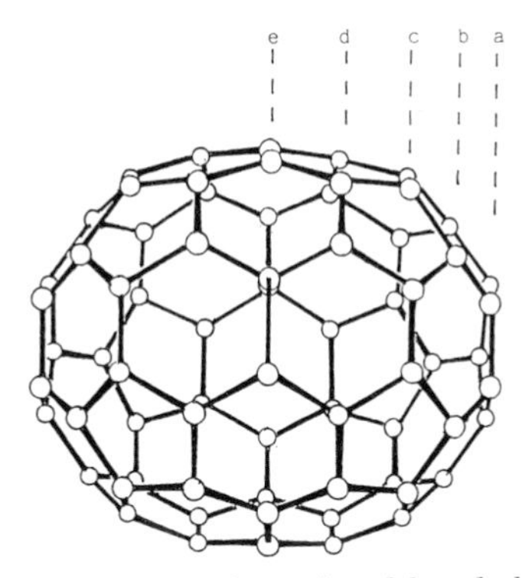

Figure 5. Bond notation for C_{70}: the c–c, a–b, and e–d bonds have the highest π-bond order.

groups in the process and giving η^6 coordination. Instead, only one group was displaced indicating that coordination to a single C_{60} bond had occurred in the manner of an electron-deficient alkene (Fagan *et al.* 1991*a*).

Addition could be expected to take place across the high-order (inter-pentagonal) bonds, which are all equivalent in C_{60}. In C_{70} the high order bonds, c–c, a–b, and e–d bonds (figure 5), have calculated (Hückel) π-bond orders of 0.602, 0.597 and 0.545 respectively (P. J. Knowles & R. Taylor, unpublished results); *ab initio* calculations give comparable results (Baker *et al.* 1991). There is, however, a second factor that needs to be taken into account, which is the eclipsing that results when two groups are adjacent, resulting in steric strain. Most organic molecules are able to twist to relieve this strain, and for those that cannot (e.g. alkenes) the *cis* groups are attached to sp^2-hybridized carbons and are thus further apart. In fullerenes, however, the rigidity of the cage severely restricts bond rotation necessary to reduce eclipsing interactions. As a result, attachment of bulky groups follows a different pattern to that involving either small groups or those with bridged groups (for which steric interactions do not apply). Both patterns of addition to C_{60} are now evident.

Attempts have been made to predict possible addition patterns. One approach considers the predominance of the decacyclene moiety in stable fullerenes, leading to the possibility that addition in which this moiety is retained would produce stable

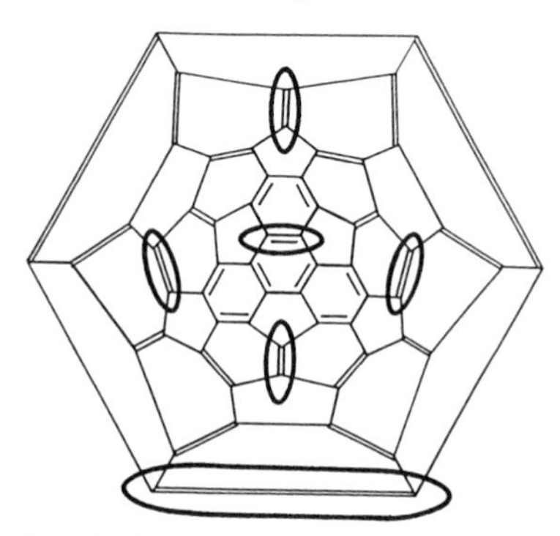

Figure 6. Octahedral addition sites (encircled) for C_{60}.

products (Taylor 1992*b*). Two addition patterns of $3X_2$ are then possible; further addition of $3X_2$ to each product in a symmetrical way leads to two further products, one of which is the octahedral derivative (figure 6); an example of the latter has now been identified (see §2*a*(iv)). This approach also indicates that some $C_{78}X_6$ compounds might be especially stable since two decacyclene moieties can be retained therein.

A second method involves calculating the stabilities of 1,2- and 1,4-addition products (figure 7). Under conditions of low steric requirement, 1,2-addition will be preferred, because location of double bonds in pentagonal rings is unfavourable. However bulky reagents will sterically interact when *cis* to each other, so 1,4-addition is then preferred. Hydrogen and fluorine are thus predicted to favour 1,2-addition, whereas chlorine, bromine and iodine will favour 1,4-addition (Dixon *et al.* 1992). Preliminary results confirm these predictions (§§2 and 3).

The strain introduced in the addition compounds causes them to readily eliminate and revert back to the parent fullerene. This makes mass spectroscopic identification of derivatives particularly difficult. A further difficulty encountered in the chemistry of fullerenes lies in the multitude of addition products that can, in principle, be obtained. For example, for C_{60} the number of products *discounting isomers* that can be produced on addition of a reagent X_2 will equal the number of 'double' bonds, i.e. 30. If isomers are included then many hundreds are possible. Development of C_{60} chemistry is therefore very much more difficult than that of benzene. It is also complicated by the low solubility of C_{60} in organic solvents, and by the fact that derivatives are themselves mostly more soluble, which facilitates polysubstitution; hydroxy and amino derivatives of C_{60} are water-soluble. The additions are described under the two main categories outlined above.

2. Addition of reagents with low steric requirements

Two types of addition are considered here: those involving a single moiety bonded to each of the two adjacent carbon atoms, and those involving two small atoms or groups.

(*a*) *Moieties bonded to adjacent cage carbons*

(i) *Oxygen*

Early work with C_{60} indicated that oxygen could be readily attached, and the observed photochemical degradation of C_{60} in benzene (Taylor *et al.* 1991) has been shown (Creegan *et al.* 1992) to produce $C_{60}O$. This compound has the epoxide (figure

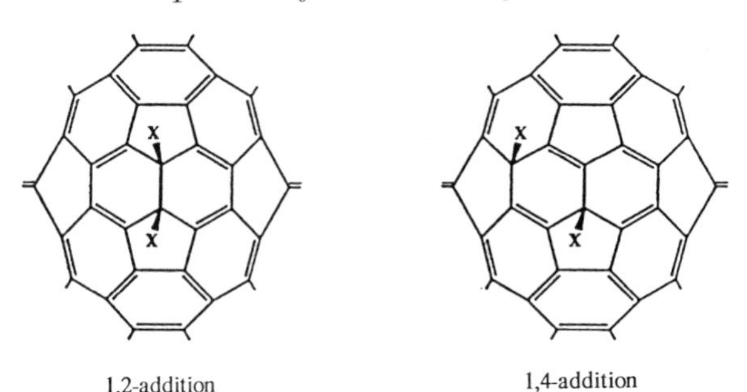

Figure 7. Products of either 1,2- or 1,4-addition to C_{60}.

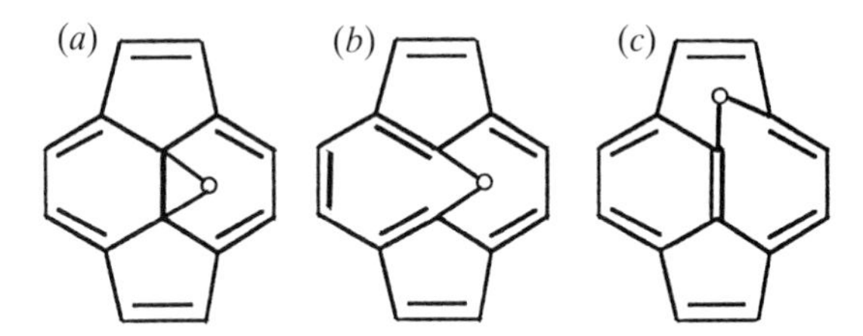

Figure 8. Conceivable structures from addition of oxygen to C_{60}.

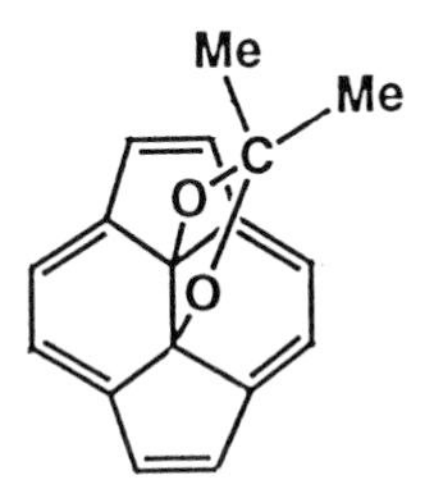

Figure 9. 1,3-Dioxolane derivative of C_{60}.

8*a*) rather than the oxidoannulene structure (figure 8*b*), the preference for avoiding inclusion of double bonds in the pentagonal rings evidently being important here (see §2*a*(v)). Recent calculations indicate that the most stable structure should in fact be the oxidoannulene (figure 8*c*) in which oxygen occupies a 5,6-ring junction (Raghavachari 1992). The advantage of this structure is that it retains the benzenoid character of one of the hexagonal rings, and models show it is considerably less strained. The comparable structure resulting from carbene addition has recently been identified (§2*a*(v)).

The epoxide is also formed from the reaction of C_{60} with dimethyldioxirane, and the 1,3-dioxolane (figure 9) is also obtained (Elemes *et al.* 1992). Photochemical irradiation of C_{60} in hexane had previously been shown to result in the addition of 2–5 oxygens, and also various methylene groups (the results depend upon the fullerene purity) (Wood *et al.* 1991). In this work, two oxygens were found to add to C_{70}.

(ii) *Osmium tetroxide*

Osmium tetroxide adds to C_{60} in the presence of pyridine to give an osmate ester (Hawkins *et al.* 1990). Replacement of pyridine by 4-*t*-butylpyridine yields material suitable for single crystal X-ray spectroscopy (Hawkins *et al.* 1991; Hawkins 1992*a*) which shows that addition has occurred across one of the high-order bonds (figure

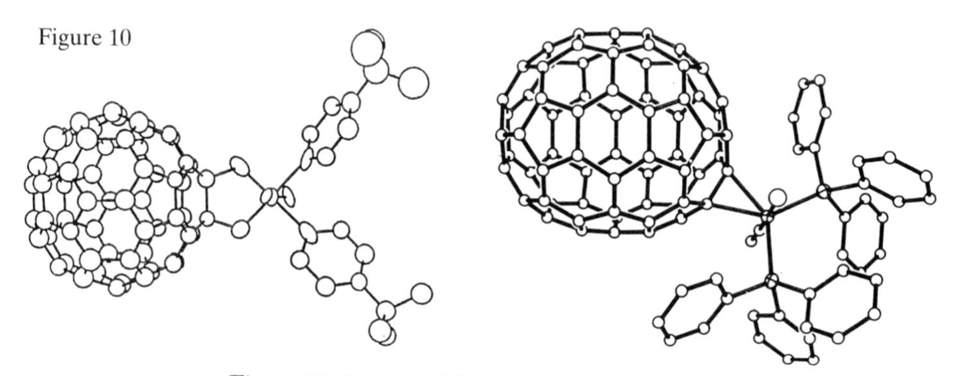

Figure 10. Adduct of C_{60} with osmium tetroxide.
Figure 11. Adduct of C_{70} with $Ir(CO)Cl(PPh_3)_2$.

10). Bis-osmylation of C_{60} is feasible but the positions of each of the second osmate groups has yet to be determined (Hawkins 1992*b*). If octahedral locations (figure 6) are involved, two disubstituted and three trisubstituted derivatives are possible. These intermediates can readily be converted into hydroxy compounds and hence, in principle, into polymers by reaction with chlorides of dicarboxylic acids.

(iii) *Iridium*

The 6,6-ring fusion site is also involved in reaction with $Ir(CO)Cl(PPh_3)_2$ to give $(\eta^2\text{-}C_{60})Ir(CO)Cl(PPh_3)_2.5C_6H_6$ (Balch *et al.* 1991*a*); the reagent here is known to give stable η^2 adducts with electron-deficient alkenes (Vaska 1968). The same reagent reacts with the a–b bond in C_{70} (figure 11). The c–c bond has the slightly higher bond order (see §1), but steric considerations may favour formation of the observed product (Balch *et al.* 1991*b*). An iridium complex in which one of the phenyl groups on phosphorus was replaced by *p*-benzyloxybenzyl gave a derivative in which the phenyl groups chelated the fullerene cage of the adjacent molecule giving rise to an infinite chain of ordered molecules (Balch *et al.* 1992). A further iridium complex of C_{60}, $(\eta^5\text{-indenyl})(CO)Ir(\eta^2\text{-}C_{60})$ has been prepared (Koefod *et al.* 1991).

(iv) *Platinum, palladium, iron, and nickel*

The addition of platinum provides information on multiple additions to C_{60}. Reaction with $(Ph_3P)_2Pt(\eta^2\text{-}C_2H_4)$ resulted in single-fold addition to give $(Ph_3P)_2Pt\eta^2\text{-}C_{60}$ (Fagan *et al.* 1991*a*), and similar derivatives have been made using nickel and palladium derivatives (Fagan *et al.* 1992). The corresponding triethylphosphine platinum compound, gave six groups coordinated to C_{60} in an octahedral manner (figure 12) (Fagan *et al.* 1991*b*). Addition of two platinum groups produces three isomers, one of which is the expected 1,9-*trans* isomer (figure 13). However, the others were deduced to be either the 1,4-, 1,6-, 1,7-, or 1,8-isomers, none of which is on the pathway to octahedral addition, suggesting that adduct migration must accompany reaction.

Reaction of C_{60} with the complex $Pd_2(\text{dibenylideneacetone})_3.CHCl_3$ results in displacement of the ligand to give $C_{60}Pd_n$. The value of n is increased if larger stoicheiometric amounts of the complex are used, but was never less than 1.0 when excess C_{60} was used. On heating the complex $C_{60}Pd$ (considered to consist of a linear

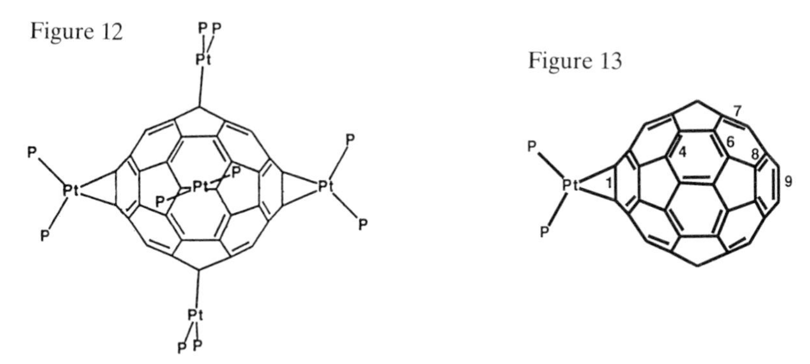

Figure 12. Octahedral co-ordination of platinum to C_{60}; the sixth platinum group is behind the cage.

Figure 13. Possible bond locations of the second group in diadducts of platinum with C_{60}.

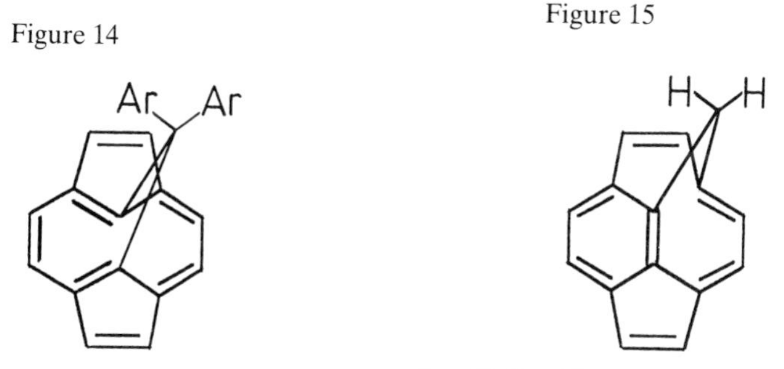

Figure 14. Bonding structure in diarylfulleroids.

Figure 15. Bonding structure in the parent fulleroid, $C_{61}H_2$.

chain of palladiums bonded to two C_{60} molecules), C_{60} is regenerated, and the value of n increases to *ca*. 3.0, indicating the formation of a three-dimensional polymer (Nagashima *et al*. 1992).

The formation of both $C_{60}Fe^+$ and $C_{70}Fe^+$ have been described (Roth *et al*. 1991; Jiao *et al*. 1992), as has the dumb-bell-shaped complex $C_{60}Ni^+C_{60}$ (Huang & Frieser 1991). In these compounds, coordination to the high π-density bonds is probable.

(v) *Diaryldiazomethane*

The reaction of aryldiazomethanes with C_{60} results (after loss of nitrogen from the initial bridged intermediate) in positioning of Ar_2C across one of the high order bonds to give 'fulleroids' (figure 14). *para*-Substituents in the aryl ring include Me, OMe, Br, Me_2N, and OCOPh, and up to six diarylcarbon groups can be attached, suggesting an octahedral array. Derivatives of fluorene can also be attached (Suzuki *et al*. 1991; Wudl 1992). A significant feature of this work is that if the aryl rings contain either hydroxy or amino groups then the possibility of producing fullerene polymers emerges. Some derivatives appear to exist in at least two isomeric forms; the predominant one is shown in figure 14, and differs from that obtained in the analogous oxygen addition (figure 8). There is evidently a fine balance between the various structural factors which favour either the open or closed bridge forms. (It has now been shown that the diarylfulleroids have the closed bridge structure analogous to that in figure 8*a* (F. Wudl, personal communication).) This is also demonstrated

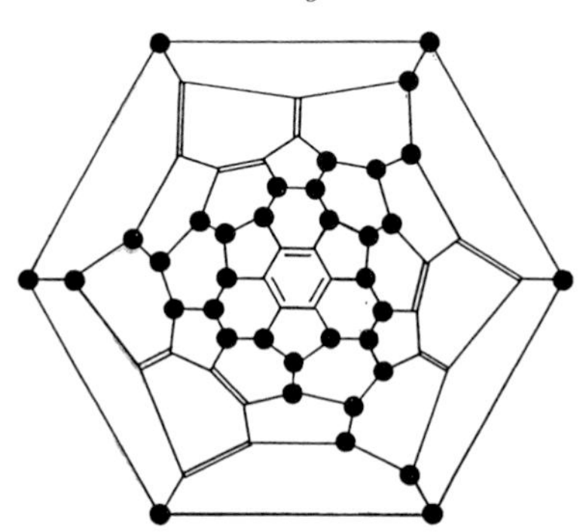

Figure 16. Possible tetrahedral structure for $C_{60}H_{36}$; filled circles denote hydrogen locations.

by the structure of the parent fulleroid (figure 15), for here a 1,5-shift appears to occur to give 6,5-bridging (Suzuki *et al.* 1992*b*). This structure has the advantage that it restores the benzenoid character of one of the hexagonal rings; the low steric requirement of hydrogen should also enhance the C–C–C angle at the expense of the H–C–H angle, thereby facilitating bridging of the greater carbon–carbon distance.

By using of *m*- and *p*-bridging groups, fulleroids with two C_{60} cages have also been prepared (Suzuki *et al.* 1992*a*).

(*b*) *Addition of separate atoms or groups*

(i) *Hydrogen*

A preliminary report of the Birch reduction (liquid ammonia) of C_{60} indicated addition of 36 hydrogens to the cage (Haufler *et al.* 1990). However, various later workers find that this reaction gives products containing amino groups (due to the ease of addition of nucleophiles to the cage). A structure proposed for $C_{60}H_{36}$, with all double bonds non-conjugated and in pentagonal rings, must be considered unlikely. A tetrahedral structure has also been proposed (Dunlap *et al.* 1991), but the most promising candidate would seem to be one with benzene rings (with little delocalization restriction) at the apices of a tetrahedron (figure 16). Mass spectrometry (CI conditions) has revealed the presence of up to 37 (Schröder *et al.* 1992) and 56 hydrogens (R. G. Cooks, personal communication) attached to C_{60}. Many adjacent sites must be occupied by hydrogen in these derivatives.

(ii) *Fluorine*

Large numbers of fluorine atoms can be added to C_{60} so they must occupy adjacent sites. The major product appears to contain *ca.* 42 fluorine atoms (Selig *et al.* 1991; Holloway *et al.* 1991), and ^{19}F NMR indicated the formation also of a small amount of $C_{60}F_{60}$ (Holloway *et al.* 1991). The fluorinated fullerenes (especially those containing many fluorines), are unstable and eliminate fluorine on standing, and the fluorine lability makes it difficult to obtain reliable mass spectra. The early notion that $C_{60}F_{60}$ would be a good lubricant neglected the strain in the molecule due to the eclipsing interactions; in PTFE such interactions are minimized by twisting of the carbon backbone, very difficult in C_{60}. The C–F bond energy is calculated to be reduced by 14% compared with CF_4 (Scuseria 1991). Some twisting of the fluorines is conceivable so that the symmetry becomes I (the molecule would then be chiral) rather than I_h (Fowler *et al.* 1991), though calculations indicate the latter to be the more stable form (Scuseria & Odom 1992).

The fluorines in fluorinated C_{60} can be displaced by a variety of nucleophiles. Furthermore, fluorinated C_{60} derivatives are likely to be synthetically useful because they are both more reactive, and more soluble than other halogenated derivatives (Taylor *et al.* 1992*b*). They thus react readily with water, which negates their use as lubricants (Taylor *et al.* 1992*a*). Mass spectrometry indicates that the products contain numerous hydroxy groups and epoxide links (probably from elimination of either HF or H_2O from adjacent groups). An addition–elimination mechanism is probably involved since the normal S_N2 process is ruled out because backside attack (Taylor *et al.* 1992*b*) is impossible.

(iii) *Nucleophiles*

C_{60} reacts with various nucleophiles (Wudl 1992), and both C_{60}Ht-Bu and C_{60}HEt have been isolated; the addition of up to six *t*-butyl groups has been detected (Hirsch *et al.* 1992). The water-soluble derivatives $C_{60}H_n(NRR6)_n$, (n is mainly 6, but up to 12 in minor components) have been obtained by reacting amines with C_{60}. The hydrogens on the cage appeared to be undergoing a series of 1,5-sigmatropic shifts, even though this places two double bonds in the pentagonal rings (Hirsch *et al.* 1991).

3. Addition of bulky reagents

(i) *Bromine and chlorine*

Chlorination of C_{60} results in the uptake of either twelve (Tebbe *et al.* 1991) or approximately 24 chlorine atoms (Olah *et al.* 1991*a*). This latter number is particularly significant in view of subsequent bromination results. Like the fluorofullerenes, the halogen in chlorofullerenes can be replaced by methoxy groups and up to 28 groups were detected (Olah *et al.* 1991*a*). This result may not, however, reflect the number of chlorines present since nucleophilic attack on C_{60} itself is possible. On heating the chloro compounds C_{60} is regenerated; likewise no mass spectrometric evidence for the chloro compounds could be obtained because the halogen is labile.

Treatment of C_{60} with neat bromine produces a compound which contains about twenty-eight bromine atoms fullerene cage (Tebbe *et al.* 1991; Birkett *et al.* 1992). The clue as to where these bromines were located came with the discovery that on bromination in selected solvents, the compounds $C_{60}Br_6$ and $C_{60}Br_8$ are obtained. Each of these derivatives, the structures of which were determined by single crystal X-ray spectroscopy (figures 17 and 18), contains approximately two molecules of bromine solvate per cage (Birkett *et al.* 1992). This suggested that the fully brominated material was similarly solvated $C_{60}Br_{24}$, with the bonded bromines spread over the surface in the same relative dispositions as in $C_{60}Br_8$ (figure 19) (Birkett *et al.* 1992). This conjecture was subsequently proved to be correct (Tebbe *et al.* 1992). Each of the bromo derivatives gives C_{60} on heating, and defies detection by mass spectrometry; the Br_6 compound can be converted into the Br_8 compound on heating in a suitable solvent, and a possible mechanism for this rearrangement has been proposed (Birkett *et al.* 1992).

(ii) *Phenyl and methyl groups*

Reaction of bromine, $FeCl_3$, benzene and C_{60} was reported to give attachment, in an unspecified way, of up to six *benzene* rings to C_{60} (Hoke *et al.* 1991). Repetition of this work but with heating of the reagents and work up followed by mass

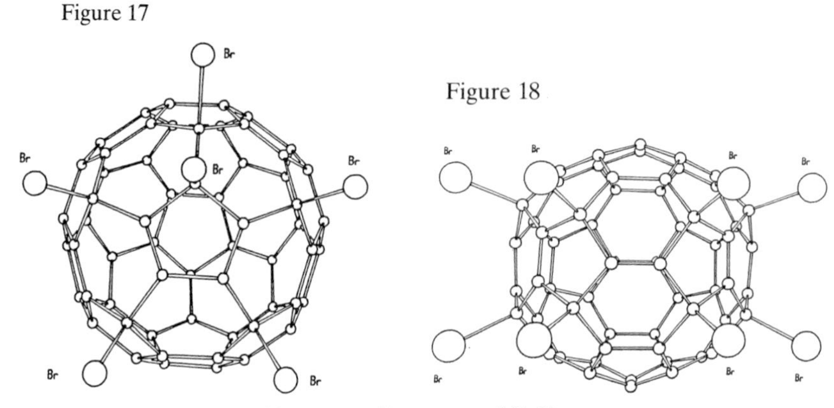

Figure 17. Structure of $C_{60}Br_6$.
Figure 18. Structure of $C_{60}Br_8$.

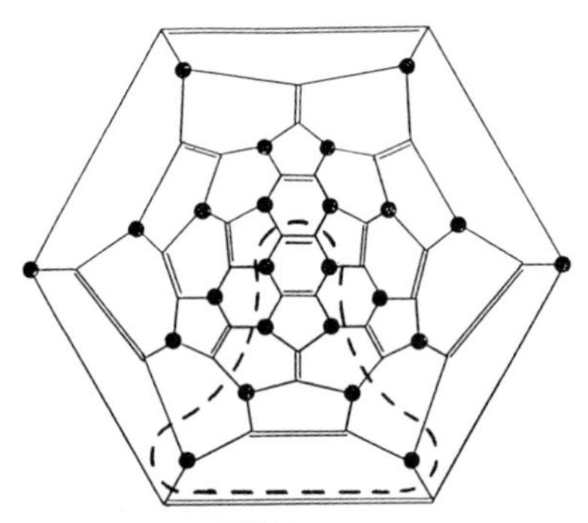

Figure 19. Schlegel diagram for $C_{60}Br_{24}$, where filled circles denote bromine locations; enclosed area shows a $C_{60}Br_8$ subset.

spectrometry showed that up to 16 *phenyl* groups are attached to the cage, ions attributable to species containing six and eight (and to a lesser extent twelve) phenyls being the most intense (Taylor *et al.* 1992*c*). The mechanism must involve electrophilic substitution of benzene by the preformed bromofullerenes, a view supported by the subsequent isolation of some of the corresponding bromofullerenes (§3(i)), which also suggests the location sites of the phenyl groups. Moreover, on reaction of C_{60} anions with methyl iodide, the dominant products also contain six and eight methyl groups (Bausch *et al.* 1991); species with up to 24 methyls were also detected. The reaction of polychlorinated C_{60} with benzene/$AlCl_3$ revealed addition of up to 22 phenyl groups. The reaction of C_{60} with benzene/$AlCl_3$ also indicated the addition of 12 (and to a lesser extent 16) H–Ph groups (Olah *et al.* 1991*b*).

4. Addition of radicals

Alkyl radicals react with C_{60} to give stable products (Krusic *et al.* 1991*b*) and the allylic $R_3C_{60}^{\cdot}$ and cyclopentadienyl $R_5C_{60}^{\cdot}$ (figure 20) radicals have been identified; up to 34 methyl groups have been added to C_{60} (Krusic *et al.* 1991*a*). The structure of the cyclopentadienyl radical suggests that a related intermediate may be involved in the formation of $C_{60}Br_6$. More recently, $RC_{60}^{\cdot}$ radicals have been prepared, and the unpaired electron shown to be confined mainly to the 6:6 ring fusion; extensive

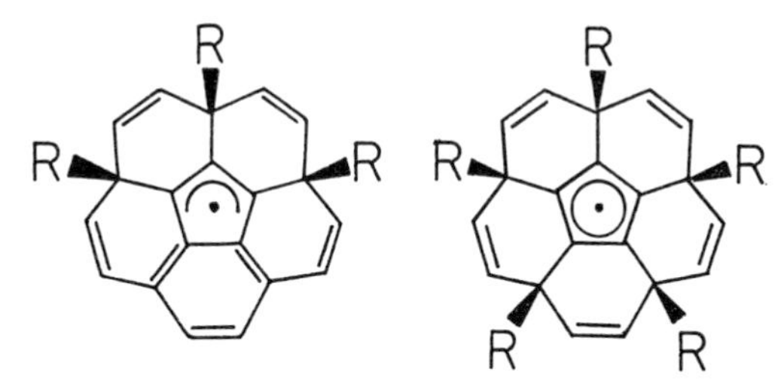

Figure 20. Structures of allylic and cyclopentadienyl radicals formed from C_{60}.

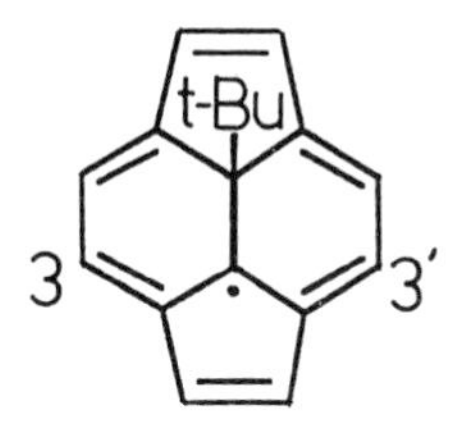

Figure 21. Probable dimerization sites for t-BuC_{60}.

delocalization is absent (Morton *et al.* 1992*b*). The radical signal from t-$BuC_{60}^{\bullet}$ increases with increasing temperature and vice versa, and the cycle can be repeated several times without significant radical decay. Dissociation of a radical dimer is indicated, and since there is little extensive radical delocalization, head-to-head dimerization is probable, most probably involving the 3- and 3′-sites (figure 21) to minimize steric interactions (Morton *et al.* 1992*b*).

A C_{60}-*p*-xylylene copolymer (-$[C_{60}]_n$-$[p\text{-xylylene}]_m$-$)_x$, $m/n = 3.4$, has been prepared by reacting the xylylene diradical with C_{60}. The product is thought to be cross-linked, but is not air stable (Loy & Assink 1992).

References

Baker, J., Fowler, P. F., Lazzaretti, J. L., Malagoli, M. & Zanasi, R. 1991 Structure and properties of C_{70}. *Chem. Phys. Lett.* **184**, 182–185.

Balch, A. L., Catalano, V. J. & Lee, J. W. 1991*a* Preparation and structure of $(\eta^2\text{-}C_{60})$Ir-(CO)Cl$(PPh_3)_2 . 5C_6H_6$. *Inorg. Chem.* **30**, 3980–3981.

Balch, A. L., Catalano, V. J., Lee, J. W., Olmstead, M. M. & Parkin, S. R. 1991*b* $(\eta^2\text{-}C_{70})$Ir-(CO)Cl$(PPh_3)_2$: The synthesis and structure of an organometallic derivative of a higher fullerene. *J. Am. chem. Soc.* **113**, 8953–8955.

Balch, A. L., Catalano, V. J., Lee, J. W. & Olmstead, M. M. 1992 Supramolecular aggregation of an $(\eta^2\text{-}C_{60})$ iridium complex involving phenyl chelation of the fullerene. *J. Am. chem. Soc.* **114**, 5455–5457.

Barth, W. E. & Lawton, R. G. 1971 The synthesis of corannulene. *J. Am. chem. Soc.* **93**, 1730–1745.

Bausch, J. W. *et al.* 1991 Diamagnetic polyanions of the C_{60} and C_{70} fullerenes: Preparation, ^{13}C and ^{7}Li NMR spectroscopic observation, and alkylation with methyl iodide. *J. Am. chem. Soc.* **113**, 3205–3206.

Bent, H. A. 1961 An appraisal of valence bond structures and hybridisation in compounds of first row elements. *Chem. Rev.* **61**, 275–311.

Birkett, P. R. *et al.* 1992 Preparation and characterization of $C_{60}Br_6$ and $C_{60}Br_8$. *Nature, Lond.* **357**, 479–481.

Bochvar, D. A. & G'alpern, E. G. 1973 Carbododecahedron, s-icosahedrane and carbo-s-icosahedron (C_{60}). *Proc. Acad. Sci. USSR* **209**, 239–241.

Creegan, K. M. *et al.* 1992 Synthesis and characterization of $C_{60}O$. *J. Am. chem. Soc.* **114**, 1103–1105.

David, W. I. F. *et al.* 1991 Crystal structure and bonding of ordered C_{60}. *Nature, Lond.* **353**, 156–158.

Dixon, D. A., Matsuzawa, N., Fukunaga, T. & Tebbe, F. N. 1992 Patterns of addition to C_{60}. *J. phys. Chem.* **96**, 6107–6110.

Dunlap, B. I., Brenner, D. W., Mintmire, J. W., Mowrey, R. C. & White, C. T. 1991 Geometric and electronic structures of $C_{60}H_{60}$, $C_{60}F_{60}$ and $C_{60}H_{36}$. *J. phys. Chem.* **95**, 5763–5768.

Elemes, Y. *et al.* 1992 Reaction of C_{60} with dimethyldioxirane – formation of an epoxide and a 1,3-dioxolane derivative. *Angew. Chem. Int. Edn. Engl.* **31**, 351–353.

Fagan, P. J., Calabrese, J. C. & Malone, B. 1991*a* The chemical nature of C_{60} and the characterization of a platinum derivative. *Science, Wash.* **252**, 1160–1161.

Fagan, P. J., Calabrese, J. C. & Malone, B. 1991*b* Multiply substituted C_{60} with an octahedral array of platinum atoms. *J. Am. chem. Soc.* **113**, 9408.

Fagan, P. J., Calabrese, J. C. & Malone, B. 1992 Metal complexes of C_{60}. *Acc. Chem. Res.* **25**, 134–142.

Fowler, P. F. 1992 Localised models and leapfrog structures of fullerenes. *J. chem. Soc. Faraday Trans.* **88**, 145–146.

Fowler, P. F., Kroto, H. W., Taylor, R. & Walton, D. R. M. 1991 Hypothetical twisted structure for $C_{60}F_{60}$. *J. chem. Soc. Faraday Trans.* **87**, 2685–2686.

Haufler, R. E. *et al.* 1990 Efficient production of C_{60}, $C_{60}H_{36}$, and the solvated buckide ion. *J. phys. Chem.* **94**, 8634–8636.

Hawkins, J. M. 1992*a* Osmylation of C_{60}: proof and characterization of the soccer-ball framework. *Acc. Chem. Res.* **25**, 150–156.

Hawkins, J. M. 1992*b* Recent results in the synthesis and characterization of discrete derivatives of C_{60}. *J. electrochem. Soc.* **139**, 241c.

Hawkins, J. M. *et al.* 1990 Organic chemistry of C_{60}: chromatography and osmylation. *J. org. Chem.* **55**, 6250–6252.

Hawkins, J. M., Meyer, A., Lewis, T. A., Loren, S. D. & Hollander, F. J. 1991 Crystal structure of osmylated C_{60}. *Science, Wash.* **252**, 312–314.

Hirsch, A., Li, Q. & Wudl, F. 1991 Globe-trotting hydrogens on the surface of the fullerene compound $C_{60}H_6[N(CH_2CH_2)_2O]_6$. *Angew. Chem. Int. Edn Engl.* **30**, 1309–1310.

Hirsch, A., Soi, A. & Karfunkel, H. R. 1992 Titration of C_{60}: a method for the synthesis of organofullerenes. *Angew. Chem. Int. Edn Engl.* **31**, 766–768.

Hoke, S. H., Molstad, J., Payne, G. L., Kahr, B., Ben-Amotz, D. & Cooks, R. G. 1991 Aromatic hydrocarbon derivatives of fullerenes. *Rapid Commun. Mass Spectrosc.* **5**, 472–474.

Holloway, J. H. *et al.* 1991 Fluorination of buckminsterfullerene. *J. chem. Soc. Chem. Commun.* 966–969.

Huang, Y. & Freiser, B. S. 1991 Synthesis of $Ni(C_{60})_2^+$ in the gas phase. *J. Am. chem. Soc.* **113**, 8186–8187.

Jiao, D. *et al.* 1992 The unique gas-phase reactivity of C_{60}^+ and C_{70}^+ with $Fe(CO)_5$. *J. Am. chem. Soc.* **114**, 2726–2727.

Klein, D. J., Schmalz, T. G., Hite, T. G. & Seitz, W. A. 1986 Resonance in buckminsterfullerene. *J. Am. chem. Soc.* **108**, 1301–1302.

Koefod, R. S., Hudgens, M. F. & Shapley, J. R. 1991 Preparation and properties of an indenyliridium(I) complex. *J. Am. chem. Soc.* **113**, 8957–8958.

Kroto, H. W. 1987 The stability of the fullerenes C_n (n = 28, 32, 36, 50, 60, and 70). *Nature, Lond.* **42**, 529–531.

Krusic, P. J., Wasserman, E., Keizer, P. N., Morton, J. R. & Preston, K. F. 1991*a* Radical reactions of C_{60}. *Science, Wash.* **254**, 1183–1185.

Krusic, P. J. *et al.* 1991*b* ESR study of the radical reactivity of C_{60}. *J. Am. chem. Soc.* **113**, 6274–6275.

Loy, D. A. & Assink, R. A. 1992 Synthesis of a C_{60}-*p*-xylylene copolymer. *J. Am. chem. Soc.* **114**, 3977–3978.

Morton, J. R., Preston, K. F., Krusic, P. J., Hill, S. A. & Wasserman, E. 1992*a* ESR studies of the reaction of alkyl radicals with C_{60}. *J. phys. Chem.* **96**, 3576–3578.

Morton, J. R., Preston, K. F., Krusic, P. J., Hill, S. A. & Wasserman, E. 1992*b* The dimerisation of RC_{60} radicals. *J. Am. chem. Soc.* **114**, 5454–5456.

Nagashima, H., Nakaoka, A., Saito, Y., Kato, M., Kawanishi, T. & Itoh, K. 1992 $C_{60}Pd_n$: the first organometallic polymer of C_{60}. *J. chem. Soc. Chem. Commun.*, 377.

Olah, G. A. *et al.* 1991*a* Chlorination and bromination of fullerenes. Nucleophilic methoxylation of polychlorofullerenes and their $AlCl_3$-catalysed reaction with aromatics. *J. Am. chem. Soc.* **113**, 9385–9387.

Olah, G. A. *et al.* 1991*b* Polyarenefullerenes obtained by acid-catalysed fullerenation of aromatics. *J. Am. chem. Soc.* **113**, 9387–9388.

Raghavachari, K. 1992 Structure of $C_{60}O$: unexpected ground state geometry. *Chem. Phys. Lett.* **195**, 221–224.

Roth, L. M. *et al.* 1991 Evidence for an externally bound FeC_{60}^+ complex in the gas phase. *J. Am. chem. Soc.* **113**, 6298–6299.

Schmalz, T. A., Seitz, W. A., Klein, D. & Hite, G. 1986 C_{60} carbon cages. *Chem. Phys. Lett.* **130**, 203–207.

Schmalz, T. A., Seitz, W. A., Klein, D. & Hite, G. 1988 Elemental carbon cages. *J. Am. chem. Soc.* **110**, 1113–1127.

Schröder, D., Bohme, D. K., Weiske, T. & Schwartz, H. 1992 Chemical signatures of C_{60} under chemical ionisation conditions. *Int. J. Mass Spectrometry* **116**, R13–R21.

Scuseria, G. E. 1991 Theoretical predictions of the equilibrium geometries of C_{60}, $C_{60}H_{60}$, and $C_{60}F_{60}$. *Chem. Phys. Lett.* **176**, 423.

Scuseria, G. E. & Odom, G. K. 1992 Exo-fluorinated $C_{60}F_{60}$ has I_h symmetry. *Chem. Phys. Lett.* **195**, 531–533.

Selig, H. *et al.* 1991 Fluorinated fullerenes *J. Am. chem. Soc.* **113**, 5475–5476.

Suzuki, T., Li, Q., Khemani, K. C., Wudl, F. & Almarsson, O. 1991 Synthesis of diphenylfulleroids C_{61} to C_{66}. *Science, Wash.* **254**, 1186–1188.

Suzuki, T., Li, Q., Khemani, K. C., Wudl, F. & Almarsson, O. 1992*a* Synthesis of *m*-phenylene- and *p*-phenylenebis(phenylfulleroids). *J. Am. chem. Soc.*, **114**, 7300–7301.

Suzuki, T., Li, Q., Khemani, K. C., Wudl, F. & Almarsson, O. 1992*b* Dihydrofulleroids H_2C_{61}: synthesis and preparation of the parent fulleroid. *J. Am. chem. Soc.* **114**, 7301–7302.

Taylor, R. 1991 A valence bond approach to explaining fullerene stabilities. *Tetrahedron Lett.* 3731–3734.

Taylor, R. 1992*a* Rationalisation of the most stable isomer of a fullerene, C_n. *J. chem. Soc. Perkin Trans.* **2**, 3–4.

Taylor, R. 1992*b* Will fullerene compounds $C_{78}X_6$ be readily formed? *J. chem. Soc. Perkin Trans.* **2**, 1667–1669.

Taylor, R. *et al.* 1991 Degradation of C_{60} by light. *Nature, Lond.* **351**, 277.

Taylor, R. *et al.* 1992*a* No lubricants from fluorinated C_{60}. *Nature, Lond.* **355**, 27–28.

Taylor, R. *et al.* 1992*b* Nucleophilic substitution of fluorinated C_{60}. *J. chem. Soc. Chem. Commun.* 665–667.

Taylor, R. *et al.* 1992*c* Formation of $C_{60}Ph_{12}$ by electrophilic aromatic substitution. *J. chem. Soc. Chem. Commun.* 667–668.

Tebbe, F. N. *et al.* 1991 Multiple reverse chlorination of C_{60}. *J. Am. chem. Soc.* **113**, 9900–9901.

Tebbe, F. N. *et al.* 1992 Synthesis and single crystal X-ray structure of $C_{60}Br_{24}$. *Science, Wash.* **256**, 822–825.

Vaska, L. 1968 Reversible activation of covalent molecules by transition metal complexes. *Acc. Chem. Res.* **1**, 335–344.

Wood, J. M. *et al.* 1991 Oxygen and methylene adducts of C_{60} and C_{70}. *J. Am. chem. Soc.* **113**, 5907–5908.

Wudl, F. 1992 The chemical properties of C_{60}, and the birth and infancy of fulleroids. *Acc. Chem. Res.* **25**, 106–112.

Discussion

E. WASSERMAN (*DuPont Experimental Station, U.S.A.*). One of the particularly exciting things about $C_{60}Br_6$ to us was the unexpected presence of the pair of bromines on adjacent carbons (unlike $C_{60}Br_{24}$). (i) Please comment on the stability of the pairing of bromines compared with $C_{60}Br_8$? (ii) Do they dimerize?

R. TAYLOR. (i) $C_{60}Br_6$ is unstable; if you warm it in certain solvents it rearranges to $C_{60}Br_8$, and we have suggested a mechanism whereby one of the bromines in the eclipsed pair migrates by a 1–3 shift. In this way three quarters of the $C_{60}Br_8$ structure can be achieved. Of course bromine is present in the cage lattice which can add and complete the conversion. C_{60} is also produced during the recrystallization of $C_{60}Br_6$. It is more soluble and more easy to handle than $C_{60}Br_8$. (ii) The t-butylC_{60} radical dimerizes head-to-head. But the brominated C_{60} are somewhat sterically crowded because of the neighbouring bromines. So far we have no evidence for dimers. We attempted to detect an ESR radical signal during bromination, but were unsuccessful. This does not, of course, preclude a minute concentration of an intermediate.

E. WASSERMAN. If the dimer is isolated, the new bond formed between two C_{60} molecules should be very interesting; it would be either long and weak or short and strong. It may be shorter because of angle relations, but is could also be long because of conjugation.

R. TAYLOR. There may be more evidence (of dimers) in P. J. Krusic's work.

R. C. HADDON (*AT & T Bell Laboratories, U.S.A.)*. You mentioned that you have managed to add methylene to some of the higher fullerenes. Is this done via diazomethane?

R. TAYLOR. Our results are based upon byproducts from another reaction and we are retracing our steps to see at what stage it occurred. It is a mystery because we do not think methylating agents were present. Furthermore we cannot be absolutely certain that we have introduced three individual methylenes as opposed to a C_3H_6 chain. The mass spectrum shows the addition of one methylene quite clearly, so that three may have added en route to an octahedral array. Mass analysis shows a CH_2 group, not nitrogen.

H. W. KROTO (*University of Sussex, U.K.*). What is known about hydrogenated C_{60}? It is suggested that it might be responsible for astrophysical spectra, although there is a $C_{60}H_{36}$ result that appears to be suspect. What is the situation?

R. TAYLOR. We tried the Birch reduction and obtained amino peaks in the IR because amination occurs under the reaction conditions. We tried to get over this problem by hydrogenating over platinum at low pressure, initially in hexane and then in benzene. Benzene must be redistilled first because it contains di-octyl phthalate. Having adopted this precaution, a purple solution was obtained which became bright yellow after 2–3 days. Of course benzene is reduced as well to cyclohexane. Another problem arises because the (hydrogenated) product, although initially soluble in benzene, will not redissolve after removal of the solvent by evaporation. Another curious feature is that the product is insoluble in acetone and in water but dissolves in an acetone/water mixture! Reduction has certainly occurred since the product contains no C_{60}.

R. C. Haddon. What is known about the photochemistry of C_{60}?

R. Taylor. In early experiments we UV irradiated C_{60} in hexane, but the amount of decomposition was dependent upon the particular batch of C_{60}. Some decomposed completely within 4 h; others showed little decomposition after 24 h. Therefore we keep C_{60} in the dark. Others have reported that the decomposition of C_{60} may be triggered by an impurity. Indeed deliberate initiation of decomposition has been reported. We have tried halogenation in the light and obtained a product which is different from that isolated under conventional conditions. C_{70} is also photosensitive; it decomposes faster in hexane than in benzene.

Polyynes and the formation of fullerenes

By H. W. Kroto and D. R. M. Walton
School of Chemistry and Molecular Sciences, University of Sussex, Brighton BN1 9QJ, U.K.

The synthesis and microwave study of linear cyanopolyynes, HC_5N and HC_7N, in the mid-1970s was followed by the unanticipated detection of these, and longer chains (HC_9N and $HC_{11}N$), in space. To gain insight into the way in which such species and carbon clusters in general might form, an experiment was devised in 1985 to simulate conditions in carbon stars, involving the laser vaporization of graphite in a supersonic nozzle and detection of the resulting carbon species by mass spectrometry. This initiative resulted in the serendipitious discovery of an entirely new allotrope of carbon, C_{60}, named buckminsterfullerene after the inventor of the geodesic dome. Five to seven years later, C_{60} and other members of what is now know as the fullerene family have been isolated in macroscopic amounts, however, these exciting developments have tended to overshadow fundamental problems associated with the aggregation of carbon atoms in which acetylenes, and polyynes in particular, may play a key role.

---

1. Introduction

Acetylenes continue to provide a seemingly inexhaustible reservoir of novel materials, with extended conjugated systems embodied in polyynes, polyenes (polyacetylenes), enynes, cumulenes and various combinations thereof featuring prominently. Researches into conducting polymers are a case in point (Masuda & Higashimura 1984; Wegner 1981), as is the quest for natural products (Bohlmann *et al.* 1973; Jones & Thaller 1978) and their derivatives, some of which display high levels of pharmacological activity. A set of unique circumstances, which augur well for synthesis is partly responsible for this situation; notably the relatively high acidity of the alkynyl hydrogen (facilitating substitution and oxidative coupling) and the ease with which the triple bond can be induced to polymerize or participate in cycloadditions. The ubiquitous C_2 unit also acts as a focus for combustion studies and, for example, for investigations into the nature of soot. The results of such work often turn out to be wholly unexpected and this is particularly true in the way fullerenes were discovered.

2. Polyyne synthesis

Advances in preparative acetylene chemistry during the 1950s, by Jones & Whiting and by Bohlmann and their respective associates, led to viable syntheses of lower members of the polyyne series: $H(C\equiv C)_nH$ ($n = 2$–5), and to substituted higher polyynes, e.g. $t-Bu(C\equiv C)_{10}Bu-t$ (Jones *et al.* 1960). This work paved the way to the identification of many naturally occurring substances containing conjugated acetylenes. The approach to the highly unstable parent polyynes often involved dehydrohalogenations in liquid ammonia; a technique which although now

Phil. Trans. R. Soc. Lond. A (1993) **343**, 103–112

Printed in Great Britain

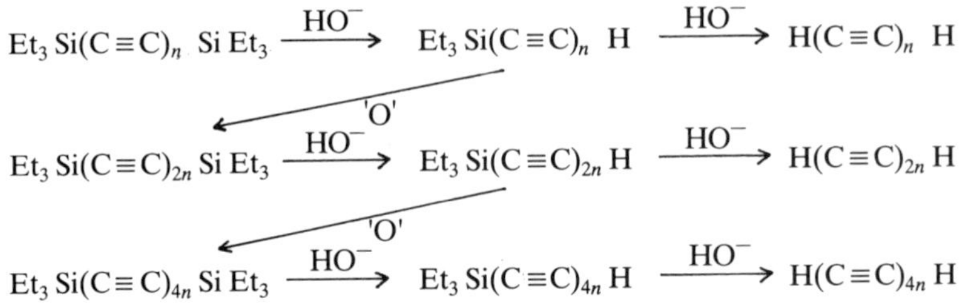

$$Et_3Si(C\equiv C)_nSiEt_3 \xrightarrow{HO^-} Et_3Si(C\equiv C)_nH \xrightarrow{HO^-} H(C\equiv C)_nH$$

$$\text{'O'}$$

$$Et_3Si(C\equiv C)_{2n}SiEt_3 \xrightarrow{HO^-} Et_3Si(C\equiv C)_{2n}H \xrightarrow{HO^-} H(C\equiv C)_{2n}H$$

$$\text{'O'}$$

$$Et_3Si(C\equiv C)_{4n}SiEt_3 \xrightarrow{HO^-} Et_3Si(C\equiv C)_{4n}H \xrightarrow{HO^-} H(C\equiv C)_{4n}H$$

Figure 1. Preparative sequence for polyynes, $H(C\equiv C)_xH$. Oxidative coupling conditions 'O' Cu_2Cl_2/TMEDA/O_2 (Eastmond *et al.* 1972).

Figure 2. Structure of $Me_3Si(C\equiv C)_4SiMe_3$ showing the curved polyyne chain arising from crystal packing forces (Coles *et al.* 1985). (Reprinted with permission of the Royal Society of Chemistry.)

well established (Brandsma 1988) is troublesome, not without hazard, and inefficient for $n \geqslant 4$. Furthermore, the range of functional groups that can be attached to the polyyne chain is limited. To circumvent these difficulties, alternative strategies were devised at Sussex during the late 1960s, based upon the capacity of silicon to protect one end of an acetylene chain during oxidative coupling of the other end (Eastmond & Walton 1968). Carefully controlled removal of one silyl group from the resulting bissilylpolyyne, followed by oxidative coupling provided a method for rapidly increasing the length of the polyyne so that species such as $Et_3Si(C\equiv C)_{16}SiEt_3$ and $H(C\equiv C)_{12}H$ could be prepared (figure 1 (Eastmond *et al.* 1972)). An attractive feature of this tactical approach lies in the fact that silyl groups confer a degree of stability on the polyynes, probably by holding the rod-like chains apart so that they cannot cross-link by highly exothermic electrocyclic processes. It is, however, worth noting that crystal lattice forces may lead to curvature in the polyyne chain (figure 2, (Coles *et al.* 1975)).

3. BSc degree by thesis

At Sussex, in the 1970s, an entirely new approach to undergraduate education – the Chemistry by Thesis degree – was initiated by C. Eaborn. Under the scheme, students attended lectures but were assessed primarily on the results of a research programme, and to avoid too narrow a focus, projects were devised by staff based in more than one chemistry discipline. The combination – synthesis plus spectroscopy – proved to be particularly fruitful in this context. A. J. Alexander, one of the first students to participate, was assigned the task of preparing cyanobutadiyne, $H(C\equiv C)_2CN$ (figure 3*a*) and measuring its microwave spectrum (Alexander *et al.* 1976). This synthesis illustrates the usefulness of the methodology referred to above, in that one metal group (in this case tin) can be selectively removed by electrophiles other than the proton, thus developing acetylene functionality.

(*a*)

$$Me_3Si(C{\equiv}C)_2SnEt_3 \xrightarrow{ClCN} Me_3Si(C{\equiv}C)_2CN \xrightarrow{Al_2O_3/H_2O} H(C{\equiv}C)_2CN$$

(*b*)

$$Et_3Sn(C{\equiv}C)_3SnEt_3 \xrightarrow{ClCN} Et_3Sn(C{\equiv}C)_3CN \xrightarrow{Al_2O_3/H_2O} H(C{\equiv}C)_3CN$$

Figure 3. Preparative schemes for $H(C{\equiv}C)_2CN$ and $H(C{\equiv}C)_3CN$.

4. Interstellar carbon chains

Alexander's project coincided with advances in molecular radioastronomy in which species such as HCO^+ were detected, notably in the black clouds found in the Milky Way galaxy. Microwave spectroscopy played a key role in this endeavour, cyanoethyne ($HC{\equiv}CC{\equiv}N$) being identified (Turner 1971) early on. In the light of this result, a collaboration with T. Oka and astronomers at the NRC, Canada, was initiated with a view to conducting a search for $H(C{\equiv}C)_2CN$ utilizing the 46 m radiotelescope located at Algonquin Park, Ontario. This quest was successful (Avery *et al.* 1976), in that cyanobutadiyne was detected in the signal emanating from Sgr B2, a giant molecular cloud near the centre of the galaxy. The following year, C. Kirby prepared $H(C{\equiv}C)_3CN$ by a route (figure 3*b*) analogous to that used by Alexander, and recorded its microwave spectrum at Sussex (Kirby *et al.* 1980). These measurements enabled a successful radio search for this species to be made, and the all-important signal was detected in Heiles's Cloud 2 (Taurus constellation). In a moment of considerable drama, the data obtained at Sussex was transmitted to Canada by telephone and found to match exactly the oscilloscope trace associated with the vital signal (figure 4) (Kroto *et al.* 1978). This result attracted wide attention because, at the time, HC_7N was the most complex molecule unambiguously to be identified in space. Subsequently HC_9N (Broten *et al.* 1978) and $HC_{11}N$ (Bell *et al.* 1982) were detected in the same way. However, lack of funding and of personnel – an all-too-common occurrence – coupled with anticipated difficulties with the synthesis, precluded preparation of these higher cyanopolyynes by conventional laboratory methods at Sussex.

5. Fullerenes from polyynes

Details of events leading to the discovery and assignment of the truncated icosahedral (football) structure to C_{60} (Kroto *et al.* 1985), the subsequent development of fullerene cage theory, the generation of macroscopic quantities and separation of C_{60} and C_{70} in 1990, and the identification of higher fullerenes have been described elsewhere (Curl & Smalley 1991; Krätschmer & Fostiropoulos 1992; Kroto 1992). The ways in which specific fullerenes, such as C_{60}, might be generated and the relevance to soot nucleation, were considered early on (Zhang *et al.* 1986; Kroto & McKay 1987) and have been reviewed against the background of carbon cluster formation (Kroto *et al.* 1991). The situation regarding carbon clusters in general is as follows (for a detailed exposition see Weltner & Van Zee 1989). Pitzer & Clementi (1959) predicted the existence of linear carbon chains C_n ($n \leqslant 10$) and species in the C_3–C_{11} range were detected soon afterwards, by Berkowitz & Chupka (1964) who vaporized graphite with a ruby laser. The chains were considered to be of the polyyne

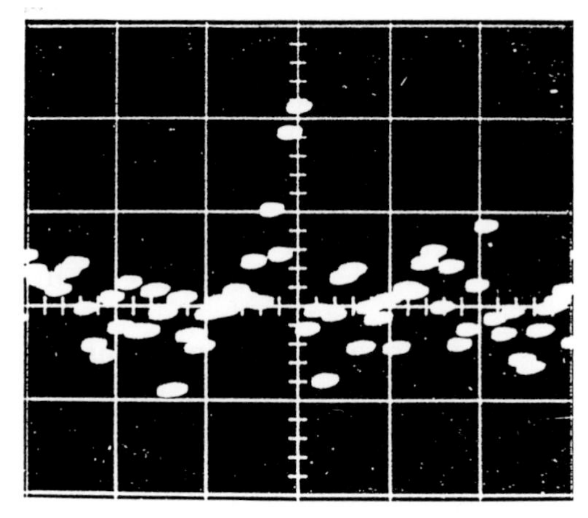

Figure 4. Radio signal (oscilloscope trace) emitted by $H(C{\equiv}C)_3CN$ in Heiles's Cloud 2.

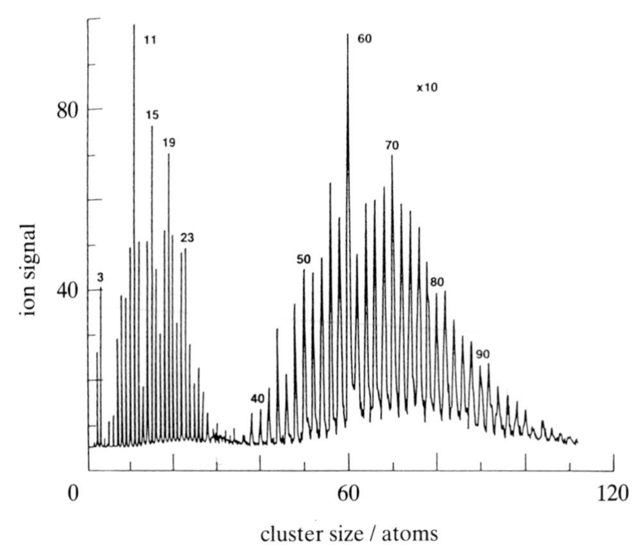

Figure 5. Mass spectrum of carbon clusters (C_3–C_{100}) generated by laser vaporization of graphite (Rohlfing *et al.* 1984). (Reprinted with permission of the American Institute of Physics.)

$-\!(C{\equiv}C)\!-_n$ or cumulene $=\!(C)\!=_n$ type. Larger clusters, C_{12}–C_{20}, were also found in which C_{11}^+, C_{15}^+ and C_{19}^+ were prominent; this phenomenon was ascribed to the fact that these species might be cyclic as the electron counts in the ions conform to the Hückel $4n+2$ rule.

Additional evidence for chain linearity ($n \leqslant 30$) was provided by the production of the cyanopolyynes upon introduction of N_2 and H_2 into the He carrier gas (Heath *et al.* 1987) and by the fact that species such as $C_{2n}K_2$ ($2 < n < 10$) were also formed when the graphite target was treated with KOH prior to vaporization (Rohlfing *et al.* 1984). These authors, using photoionization time-of-flight mass spectrometry, found only even clusters in the C_{32}–C_{100} range (figure 5), a result which they attributed to some systematic combination of C_2 units generated by the laser oblation of graphite.

Consensus of opinion now favours closed shell fullerene structures for the C_{32}–C_{100}

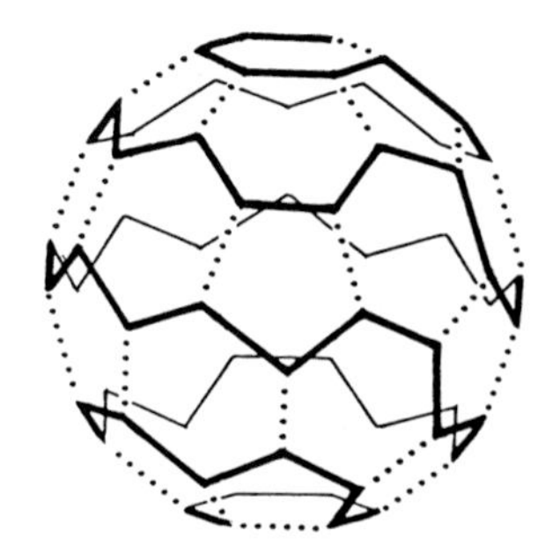

Figure 6. Spirocyclization of a C_{60} acyclic polyyne. (Reprinted with permission of VCH Publishers 1992.)

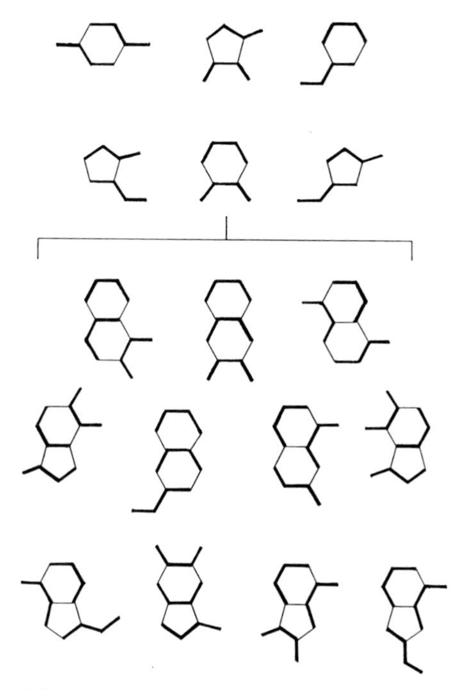

Figure 7. C_4 unit cycloadditions generating pentagon and hexagon-containing networks. (Reprinted with permission of VCH Publishers 1992.)

clusters, with each containing, by definition, the twelve pentagons required for closure (Euler's rule) plus $(n-20)/2$ hexagons. There is now convincing evidence that all fullerenes (C_{20}, C_{24} ...) may form from graphite (Kroto 1987; Hallet *et al.* 1992), however, the mechanism is very much an open question, with concrete evidence hard to come by. One possibility would be for long polyyne chains, formed from C_2 or C_4 units, to spirocyclize forming a complete or part fullerene. Figure 6 shows *in extremis* how buckminsterfullerene might be generated in this way from a 60-carbon atom polyyne. An alternative would be for C_2 or C_4 units to participate in sequential cycloadditions; figure 7 illustrates the possible conformations adopted by two C_4 units, generating a hexagon or pentagon, which in turn could act as a template for a third C_4 unit leading to fused rings (hexagon plus hexagon or hexagon plus pentagon). Inclusion of pentagons may lead to saucer-shaped closing structures

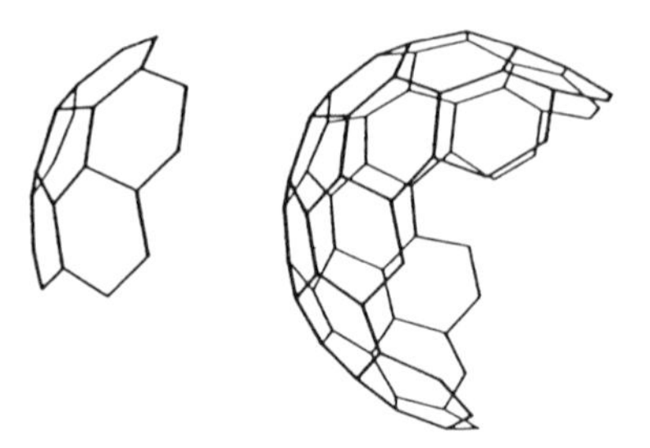

Figure 8. Saucer-shaped structures formed by inclusion of pentagons during nucleation (Kroto & McKay 1988).

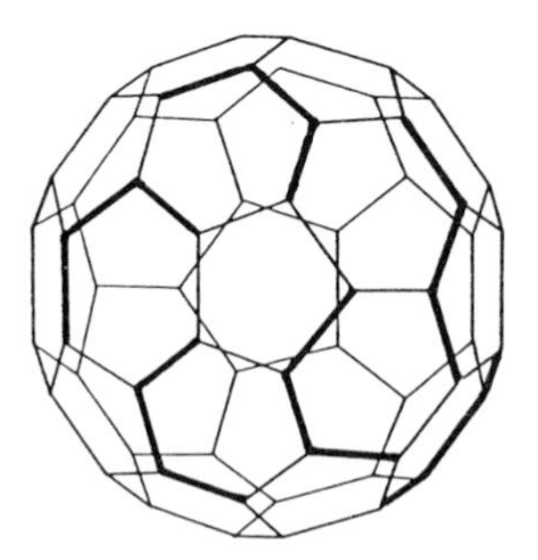

Figure 9. C_4 cycloaddition sequences leading to C_{60}. (Reprinted with permission of VCH Publishers 1992.)

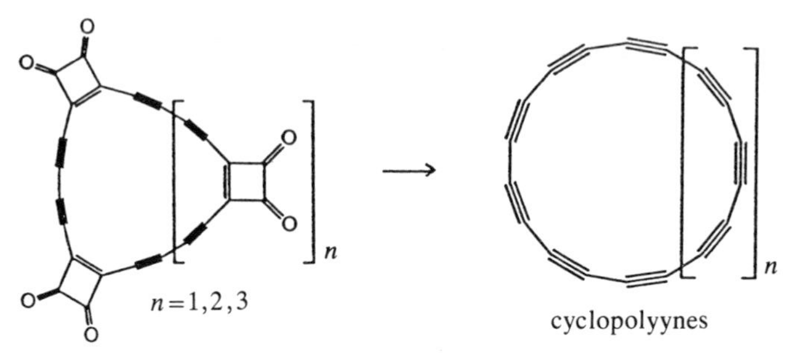

Figure 10. Generation of cyclopolyynes, C_{18} ($n = 1$), C_{24} ($n = 2$) and C_{30} ($n = 3$) by decarbonylation (Rubin *et al.* 1991). (Reprinted with permission of the American Chemical Society.)

(Kroto & McKay 1988) (figure 8), which are transients and, by a process of further nucleation, yield fullerenes. Indeed, C_{60} could in theory be built up entirely from C_4 (diyne) units by sequential cycloadditions (Kroto & Walton 1992) (figure 9).

Further insight into the process of carbon aggregation/fragmentation in the context of polyynes is provided by the remarkable results obtained by Diederich and his colleagues (Rubin *et al.* 1990) who developed elegant methods for preparing cyclic carbon oxides. When compounds $n = 1, 2, 3$ (figure 10) were submitted to negative ion laser desorption, prominent species were formed commensurate with generation of the cyclic polyynes, C_{18}^-, C_{24}^- and C_{30}^- respectively, arising from the total loss of CO fragments. The decacarbonyl, $n = 3$, also produced ions in the C_5^-–C_{33}^- range, ascribed to facile aggregation/fragmentation of the initially formed C_{30}^- species. Significantly, higher clusters were absent. In complete contrast, the positive ion desorption spectra

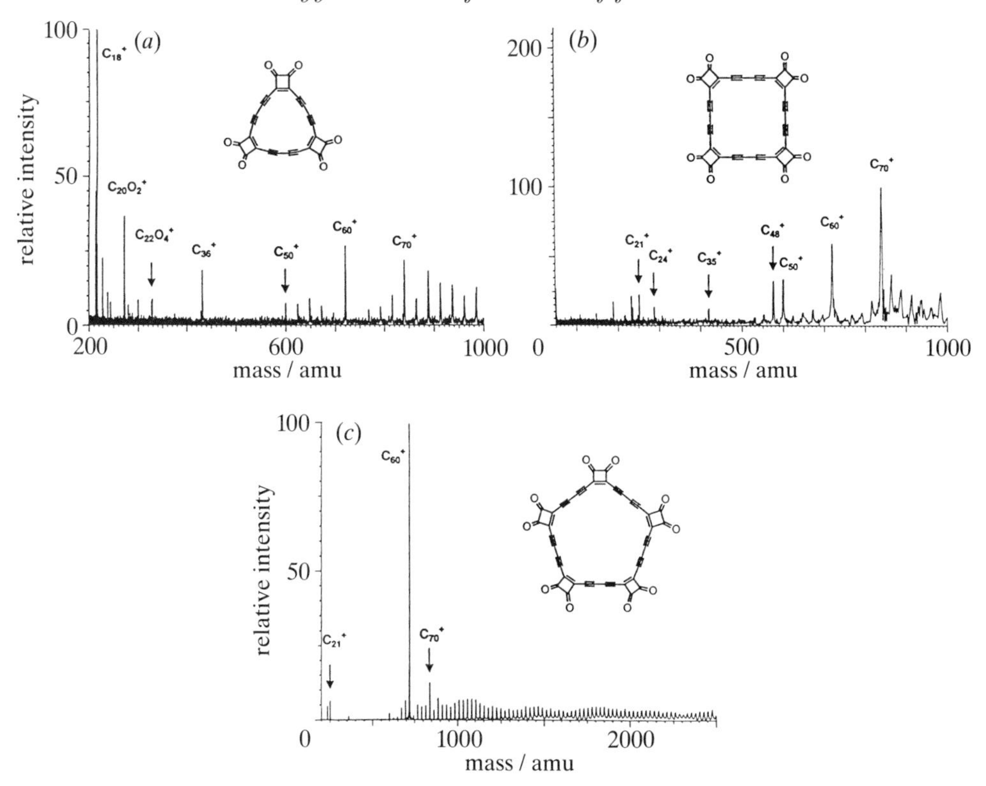

Figure 11. Positive ion laser desorption FT mass spectrum of carbon oxides (Rubin *et al.* 1991). (Reprinted with permission of the American Chemical Society.)

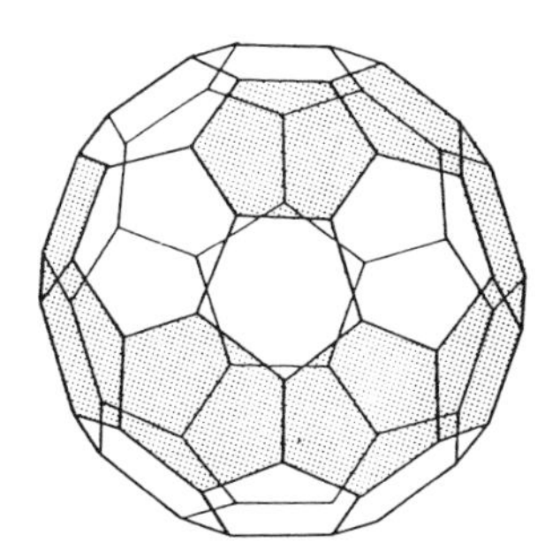

Figure 12. Spirodimerization of C_{30} polyyne rings (one C_{30} unit highlighted). (Reprinted with permission of VCH Publishers 1992.)

of $n = 1$ and 2 (figure 11) gave C_{18}^+ and C_{24}^+ together with dimers C_{36}^+ and C_{48}^+ respectively, in addition to C_{50}^+, C_{60}^+ and C_{70}^+ presumably resulting from ejection of C_2 fragments from higher carbon aggregates (multiples of C_{18} or C_{24}). Most remarkably the low-power spectrum derived from $n = 3$ exhibited a massive C_{60}^+ peak to the virtual exclusion of all other ions including C_{30}^+. This result is important in that it confirms both the relative instability of C_{30} ions, and the special stability associated with C_{60} species. A possible scheme for cross-linking within the C_{30} fragment followed by, or concomitant with, spirodimerization leading to the C_{60} ion is illustrated in figure 12. At maximum laser power all three carbon oxides generate even-numbered carbon aggregates with masses up to and in excess of C_{300}.

In conclusion we stress again that polyynes are central to carbon clusters and that this area of endeavour continues to represent a challenge when it comes to interpreting complex chemistry as epitomized by the fullerenes.

We thank British Gas, the SERC and NATO for financial support and J. W. Cornforth, F. Diederich, the late A. E. Douglas, T. Oka, R. L. Whetten and S. Wood for helpful discussions.

References

Alexander, A. J., Kroto, H. W. & Walton, D. R. M. 1976 The microwave spectrum, substitution structure and dipole moment of cyanobutadiyne, HC≡CC≡CC≡N. *J. molec. Spectrosc.* **62**, 175–180.

Avery, L. W., Broten, N. W., Macleod, J. M., Oka, T. & Kroto, H. W. 1976 Detection of the heavy interestellar molecule cyanodiacetylene. *Astrophys. J.* **205**, L173–L175.

Bell, M. B., Feldman, P. A., Kwok, S. & Matthews, H. E. 1982 Detection of $HC_{11}N$ in IRC+10°216. *Nature, Lond.* **295**, 389–391.

Berkowitz, J. & Chupka, W. A. 1964 Mass spectrometric study of vapor ejected from graphite and other solids by focused laser beams. *J. chem. Phys.* **40**, 2735–2736.

Bohlmann, F., Burkhardt, T. & Zdero, C. 1973 *Naturally occurring acetylenes.* London and New York: Academic Press.

Brandsma, L. 1988 *Preparative acetylenic chemistry*, 2nd edn. Amsterdam, Oxford, New York & Tokyo: Elsevier.

Broten, N. W., Oka, T., Avery, L. W., MacLeod, J. M. & Kroto, H. W. 1978 The detection of HC_9N in interstellar space. *Astrophys. J.* **223**, L105–L107.

Coles, B. F., Hitchcock, P. B. & Walton, D. R. M. 1985 The crystal molecular structure of 1,8-bis(trimethylsilyl)octatetrayne. *J. chem. Soc. Dalton Trans.* 442–445.

Curl, R. F. & Smalley, R. E. 1991 Fullerenes. *Scient. Am.* 54–63.

Eastmond, R. & Walton, D. R. M. 1968 The protection of terminal ethynyl groups in the oxidative couplings of acetylenes. *Chem. Commun.* 204–205.

Eastmond, R., Johnson, T. R. & Walton, D. R. M. 1972 A general synthesis of the parent polyynes $H(C{\equiv}C)_nH$ ($n = 4$–10, 12). *Tetrahedron* **28**, 4601–4616.

Hallet, R. P., McKay, J. G., Balm, S. P., Allaf, W., Kroto, H. W. & Stace, A. J. 1992 Reaction studies of carbon clusters. *J. chem. Soc. Faraday Trans.* (Submitted.)

Heath, J. R., Zhang, Q., O'Brien, S. C., Curl, R. F., Kroto, H. W. & Smalley, R. E. 1987 The formation of long carbon chain molecules during laser vaporisation of graphite. *J. Am. chem. Soc.* **109**, 359–363.

Jones, E. R. H., Lee, H. H. & Whiting, M. C. 1960 The preparation of conjugated octa- and deca-acetylenic compounds. *J. chem. Soc.* 3483–3489.

Jones, E. R. H. & Thaller, V. 1978 Natural acetylenes. In *The chemistry of the carbon—carbon triple bond* (ed. S. Patai). Chichester & New York: Wiley.

Kirby, C., Kroto, H. W. & Walton, D. R. M. 1980 The microwave spectrum of cyanohexatriyne, HC≡CC≡CC≡CCN. *J. molec. Spectrosc.* **83**, 261–265.

Krätschmer, W. & Fostiropoulos, K. 1992 Fullerite – Neue Modifikationen des Kohlenstoffs. *Physik Zeit.* **23**, 105–110.

Kroto, H. W., Kirby, C., Walton, D. R. M., Avery, L. W., Broton, N. W., MacLeod, J. M. & Oka, T. 1978 The detection of cyanohexatriyne, $H(C{\equiv}C)_3CN$, in Heiles's cloud 2. *Astrophys. J.* **219**, L133–L137.

Kroto, H. W., Heath, J. R., O'Brien, S. C., Curl, R. F. & Smalley, R. E. 1985 C_{60}: Buckminsterfullerene. *Nature, Lond.* **318**, 162–163.

Kroto, H. W. 1987 The stability of the fullerenes C_n, with $n = 24, 28, 32, 36, 50, 60$ and 70. *Nature, Lond.* **329**, 529–531.

Kroto, H. W. & McKay, K. G. 1988 The formation of quasi-icosohedral spiral shell carbon particles. *Nature, Lond.* **331**, 328–331.

Kroto, H. W., Allaf, A. W. & Balm, S. P. 1991 C_{60}: Buckminsterfullerene. *Chem. Rev.* **91**, 1213–1235.

Kroto, H. W. 1992 C_{60}: Buckminsterfullerene. The celestial sphere that fell to earth. *Angew. Chem. Int. Edn. Engl.* **31**, 111–128.

Kroto, H. W. & Walton, D. R. M. 1992 Postfullerene organic chemistry. In *Carbocyclic cage compounds* (ed. E. Osawa & O. Yonimitsu). London and New York: VCH Publishers.

Masuda, T. & Highashimura, T. 1984 Synthesis and high polymers from substituted acetylenes: exploitation of molybdenum and tungsten-based catalysis. *Acc. Chem. Res.* **17**, 51–56.

Pitzer, K. S. & Clementi, E. 1959 Large molecules in carbon vapour. *J. Am. chem. Soc.* **81**, 4477–4485.

Rohlfing, E. A., Cox, D. M. & Kaldor, A. 1984 Production and characterisation of supersonic carbon cluster beams. *J. chem. Phys.* **81**, 3322–3330.

Rubin, Y., Kahr, M., Knobler, C. B., Diederich, F. & Wilkins, C. L. 1991 Formation of cyclo[n]carbon ions C_n^+ ($n = 18, 24$) C_n^- ($n = 18, 24, 30$), and higher carbon ions including C_{60}^+ in laser desorption Fourier transform mass spectrometric experiments. *J. Am. chem. Soc.* **113**, 495–500.

Turner, B. E. 1971 Detection of interstellar cyanoacetylene. *Astrophys. J.* **163**, L35–L39.

Wegner, G. 1981 Polymers with metal-like conductivity. A review of their synthesis, structure and properties. *Angew, Chem. Int. Edn. Engl.* **20**, 361–381.

Weltner Jr, W. & Van Zee, R. J. 1989 Carbon molecules, ions, and clusters. *Chem. Rev.* **89**, 1713–1747.

Zhang, Q. L., O'Brien, S. C., Heath, J. R., Liu, Y., Curl, R. F., Kroto, H. W. & Smalley, R. E. 1986 Reactivity of large carbon clusters: spheroidal carbon shells and their possible relevance to the formation and morphology of soot. *J. phys. Chem.* **90**, 525–528.

Discussion

D. E. H. Jones (*University of Newcastle, U.K.*). You mentioned the explosive nature of polyynes. What do they explode into?

D. R. M. Walton. It is not known. They give carbon and hydrogen, but you will find very few studies, to my knowledge, in the literature of what actually happens. I think it has been such a daunting challenge, and having seen di- and tri-acetylene explode, one can understand why. But they give you carbon powder as do many acetylene compounds, quite unexpectedly. I think it is one of the supreme uninvestigated areas of science. I've not seen anything, it's been left, and the beauty is that at least it is possible to do it because we now have a method for making very pure polyynes. There is a complicating factor in that there may be the possibility of epoxidation, arising from the presence of small amounts of oxygen. But I actually believe that they decompose when absolutely pure. The challenge is to tame the explosion.

D. E. H. Jones. I do know of a report of an analysis of the explosion products of silver acetylide, and that claims to give amorphous carbon with no X-ray diffraction pattern.

H. W. Kroto. There is a fascinating experiment that has been done where it is claimed, and I think I believe it, that you can get a propagating flame in acetylene, without oxygen, and it is claimed that it goes to carbon and hydrogen.

R. C. Haddon (*AT&T Bell Laboratories, U.S.A.*). I was wondering, are there any other crystal structures, besides the one you showed, of the long polyacetylene molecules for structures known?

D. R. M. Walton. The early polyyne papers were by Jones and by Bohlmann, and to my knowledge, the crystal structures were not published. the latest papers are by Diederich and deal with cyclic carbon oxides. Don't press me for details over bond lengths, however there is a tendency towards bond equalization and the curvature is there.

Hypothetical graphite structures with negative gaussian curvature

By A. L. Mackay and H. Terrones

Department of Crystallography, Birkbeck College, University of London, Malet Street, London WC1E 7HX, U.K.

We consider the geometries of hypothetical structures, derived from a graphite net by the inclusion of rings of seven or eight bonds, which may be periodic in three dimensions. Just as the positive curvature of fullerene sheets is produced by the presence of pentagons, so negative curvature appears with a mean ring size of more than six. These structures are based on coverings of periodic minimal surfaces, and surfaces parallel to these, which are known as exactly defined mathematical objects. In the same way that the cylindrical and conical structures can be generated (geometrically) by curving flat sheets so that the perimeter of a ring can be identified with a vector in the two-dimensional planar lattice, so these structures can be related to tessellations of the hyperbolic plane. The geometry of transformations at constant curvature relates various surfaces. Some of the proposed structures, which are reviewed here, promise to have lower energies than those of the convex fullerenes.

1. Introduction

The characteristics of the process of X-ray crystal structure analysis have led to an undue emphasis on classically crystalline materials to the neglect of organized structures which do not give simple diffraction patterns with sharp spots.

Gradually, even in the inorganic field, curved layers have become recognized as essential structural components. These were first recognized in asbestos and halloysite (Whittaker 1957; Yada 1971), where concentric cylinders and spiral windings of silicate sheets were disclosed. We can now begin to assemble the basic geometry of such curved structures under the rubric of 'flexi-crystallography'. This might be part of what de Gennes (1992) has called the study of 'soft matter', the main characteristics of which are complexity and flexibility.

We assume here that we are discussing graphite layers, but most of the geometry applies to other layers, such as silicate sheets, boron nitride or boric acid, with hexagonal or square or lower symmetry. It also applies to lipid bilayers, which occur in vesicles (Fourcade *et al.* 1992) and in many biological structures. The role of liquid crystal structures as proto-organelles was recognized by Bernal (1933) and became part of his programme for generalized crystallography.

The structural components which we will chiefly consider here are the hexagonal sheets of three-connected sp^2 carbon atoms found in graphite. In graphite itself these sheets are stacked in hexagonal sequences (repeating every two sheets) or rhombohedrally (repeating every three sheets) or in disordered stacking referred to

Phil. Trans. R. Soc. Lond. A (1993) **343**, 113–127
Printed in Great Britain

as *turbostratic*. The dimensions of the hexagonal graphite structure (with two layers, 3.4 Å† apart) are: $a = 2.47$ Å, $c = 6.79$ Å so that the C–C distance is about 1.42 Å.

2. Tessellations

In a planar hexagonal lattice of lattice constant a, the distances from one lattice point to another are given by $a(h^2+hk+k^2)^{\frac{1}{2}}$ where h and k are the steps along the two hexagonal axes (which are here taken to be 60° apart: if they are taken as 120° apart, then $a(h^2-hk+k^2)^{\frac{1}{2}}$). Starting from one lattice point, a hexagonal super-lattice of side $a(h^2+hk+k^2)^{\frac{1}{2}}$ can be marked out where all its points lie on points of the original lattice. Each cell of the larger lattice will contain $(h^2+hk+k^2) = T$ of the smaller cells. This sequence $(h^2+hk+k^2) = T$ runs 1, 3, 4, 7, 9, 12, 13, ... This kind of tessellation has long been known in mineralogy where a fraction of the atoms in a hexagonal lattice may be vacant or substituted by other types of atom. The vacancies or substituting atoms are arranged symmetrically as far apart from each other as possible.

If $h > k$ and $k \neq 0$ then the super-lattice is unsymmetrically disposed with respect to the original lattice. If the lattice is turned over and superimposed on itself so that the super-lattice points coincide, then we have a coincidence site lattice (in which a fraction $1/(h^2+hk+k^2)$ of the original lattice points coincide). Coincidence site lattices can also be found in three dimensions, particularly for cubic lattices.

We will here consider curved sheets; two-dimensional manifolds. In a curved sheet, at each point there are two principal curvatures, k_1 and k_2. The mean curvature H is thus $\frac{1}{2}(k_1+k_2)$ and the gaussian curvature K is $k_1 k_2$. An ellipsoidal shell thus has positive gaussian curvature, a hyperbolic sheet has negative gaussian curvature and a cylinder or a cone has zero gaussian curvature. On a curved sheet the perimeter of a small circle of radius r is $2\pi r(1-\frac{1}{3}Kr^2+O(r^4))$. Thus for negative gaussian curvature, as on a saddle surface, there is excess area and perimeter, as compared with a plane circuit and for positive gaussian curvature, as on the sphere, the area and perimeter of a circuit are less than for a plane.

3. Cylindrical lattices

The cylinder may be developed from the plane, meaning that a sheet can be rolled up, without local distortion, to become a cylinder. Gaussian curvature is constant under such bending.

For a circular cylinder of radius r, $K_1 = 1/r$ and $K_2 = 0$, so that $H = 1/2r$ and $K = 0$. A lattice point may be marked at one identifiable point in the tessellation of a plane pattern and all identical points are then similarly marked (identical meaning also identical in orientation of surroundings). Cylindrical lattices can readily be handled by taking a cylindrical projection where the surface is unrolled to give a plane sheet of width $2\pi r$. In rational lattices further lattice points lie exactly above others with a displacement parallel to the axis of the cylinder. With irrational lattices a second lattice point never occurs exactly above the first and equalization of bond lengths tends to generate a coiled coil. The theory of diffraction from helices of both types has been developed by Klug *et al.* (1958).

The dense packings of equal spheres around a cylinder have been examined by Erickson (1973) who derived useful formulae for their generation. These can be used

† 1 Å = 10^{-10} m = 10^{-1} nm.

to produce the corresponding graphite nets by omitting the spheres at the centres of rings of six. Clearly, not all sphere packings correspond to graphite nets.

Iijima and his group at NEC (Iijima 1992; Ajayan & Iijima 1992; Iijima *et al.* 1992) have observed hollow cylinders of graphite by high-resolution electron microscopy and by electron diffraction, and have demonstrated the orientation of the lattice with respect to the axes of the cylinders. Tubes often consist of five or more layers, probably separate tubes but possibly spirals. Single layer tubes of diameter 8 Å (the diameter of the C_{60} sphere) were also seen. Conical sections were found, joining tubes of different diameters. If the cylindrical sections are flattened, then it is not necessary to postulate lines of 5–7 dislocations in the conical sections which may be seamless, the sheet edges joining at 60°. There must be at least one ring of 7 or 8 where the cone joins the smaller cylinder. Presumably successive graphite sheets in coaxial cylinders cannot be in register as in plane sheets, while their perimeters increase by $2\pi \times 3.4$ Å (about 8.7 repeat units) for each layer. There are now several studies of the electronic properties of such tubes.

Tubes of much larger diameter were earlier produced by Tibbetts *et al.* (1987) at GEC in America.

4. Regular and semi-regular polyhedra

A regular polygon is a planar polygon with all its sides of equal length, all its inter-edge angles equal and all its vertices symmetrically equivalent. If stellated, edges may intersect each other. For example the pentagram is the stellation of the pentagon in which, tracing the edges round the centre, more than one circuit is necessary to return to the starting point. The regular polygons of order 5, 8, 10 and 12 have each only one stellation, namely {5/2}, {8/3}, {10/3} and {12/5}. Other orders, such as 7, have more than one stellation. It is probably no coincidence that quasi-crystals may have symmetry axes of orders 5, 8, 10 and 12 as compared with the axes of order 2, 3, 4 and 6 (for which the corresponding polygons have no stellations) allowable in real crystals. Stellation might be considered as a first step in the generalization of the concept of axis of symmetry where coincidence occurs only after two or more rotations about the axis, recalling a Frank–Reed source.

The five regular polyhedra (the tetrahedron, cube, octahedron, dodecahedron and icosahedron) each have faces which are all the same regular polygon and vertices which are all symmetrically equivalent. The 13 semi-regular or archimedean solids are convex polyhedra which have all their faces regular polygons of two or more kinds and all vertices symmetrically equivalent. These polyhedra can be designated by the numbers of faces meeting at a vertex. For example the truncated icosahedron of C_{60} is 5.6^2. This figure is also obtained by deep truncation of the dodecahedron, truncation being, of course, a mathematical rather than a physical process. It is conventional to exclude the prisms $N.4^2$ from this definition. There is also a large number (53) of stellated regular and semi-regular polyhedra with which we will not be concerned here.

5. Euler's law

For a convex polyhedron, topologically like a sphere, with F faces, V vertices and E edges, Euler's law states that $V-E+F = 2$. A sphere has genus zero and if another more complex polyhedron can be deformed to take the shape of a sphere with N handles, then it has the genus N. This is a useful, but not a complete, characterization of shape. Knot theory, which is still developing, is needed for a better classification. A torus has the genus 1 and the P-surface inside a cubic unit cell (see below), has the genus 3. The imposition of periodic boundary conditions is equivalent to putting three handles across outside the cube connecting opposite faces. The more general expression of Euler's Law in three dimensions is $V-E+F=\chi$, where χ is the Euler characteristic and $\chi = 2-2g$, where g is the genus. Thus, for a torus, $V-E+F=0$, and for a cell of the P-surface, $V-E+F=-4$.

For a network of the graphite type, assuming that there are only pentagons, hexagons, heptagons or octagons, each edge is shared by two polygonal faces and each vertex is shared by three polygons. If N_n is the number of polygons with n sides and n vertices, we have

$$F = N_5+N_6+N_7+N_8,$$

$$2E = 3V = 5N_5+6N_6+7N_7+8N_8.$$

Putting this into Euler's expression we have

$$N_5-N_7-2N_8 = 6\chi = 12(1-g)$$

so that for a torus $N_5-N_7-2N_8=0$. For a sphere $N_5-N_7-2N_8=12$ and thus in such a figure, if there are no heptagons or octagons, there can only be hexagons, in an indefinite number, and 12 pentagons. Since $V=20+2N_6$, the number of vertices must be even. For the unit cell of the P-surface $N_5-N_7-2N_8=-24$. Thus, if there are no pentagons, 24 N_7 or 12 N_8 are necessary.

The concept of polyhedra can be extended to include infinitely periodic regular and semi-regular polyhedra by allowing non-convex arrangements. Many of these have been discussed by Wells (1977). The simplest is made up of hexagons and squares and all vertices are equivalent, each having the symbol 6.4^3, meaning that in going round a vertex we meet a hexagon and three squares. This polyhedron divides all space into two congruent regions and is a polygonal version of the P-surface.

6. Infinite polyhedra

For the convex semi-regular polyhedra the sum of the face angles meeting at a vertex adds up to less than 360° and, if there are N vertices the N deficits total 720°. The vertex sum divided into 720° is an integer or, for a stellation, a fraction.

We may ask what combinations of three faces (pentagons, hexagons, heptagons, etc.) can meet at a vertex under the condition that all vertices should be symmetrically equivalent. It can readily be seen that, if any of the faces has an odd number of sides, then the other two polygons must be the same (as two different polygons cannot alternate around an odd axis). It is more difficult to enumerate vertices at which four or more faces meet but, since we are here considering graphite sheets, this is not necessary.

For the infinite semi-regular polyhedra the vertex sums are greater than 360°, depending on the genus g, the total excess being $4\pi(g-1)$. If the polyhedron is stellated the genus may be fractional, but we are not concerned with this case.

We may consider the tessellations either as packings of pentagons, hexagons, etc. meeting three at a vertex, or as the repetitions of asymmetric triangular units. For example, the regular tessellation by heptagons with the symbol 7^3 (for three heptagons meeting at a vertex) is equivalent to the tessellation by triangles 14.6.4 of angles $\frac{1}{7}\pi$, $\frac{1}{3}\pi$, $\frac{1}{2}\pi$. By enumerating all the possibilities we find that the combinations 8.6^2, $8^2.5$, 7^3, 7.6^2 have vertex sums a little above 360°.

7. Deltahedra

A *deltahedron* is a polyhedron, each of the faces of which is an equilateral triangle. Deltahedra include the tetrahedron, the octahedron, the icosahedron and many less regular figures, such as the pentagonal bi-pyramid, the $\bar{4}$m dodecahedron, etc. Given a two-dimensional hexagonal lattice, by cutting away one, two or three sectors around a hexagonal point, it can be folded to cover a deltahedron. Thus, by cutting out sectors at the points of a larger super-lattice, a pattern of hexagonal symmetry can be mapped on to the surface of a deltahedron, so that T units lie on every face. The most important case is that of the icosahedral shells found for many viruses. Albrecht Dürer seems to have invented the construction of polyhedra by folding up cardboard and built the *icosahedron truncum* in this way. D'Arcy Thompson (1925) generalized the folding method and showed that all the convex semi-regular solids could be constructed by folding plane tilings. The corresponding process of inserting sectors to raise the order of an axis, for example, from six to seven, can be used to generate the infinite polyhedra with concavities. Possible graphite meshes can be conveniently derived from some of the dense packings of circles on a surface. Erickson (1973) has given formulae for many of the cylindrical packings and Townsend *et al.* (1992) have used this concept in decorating irregular surfaces.

8. Fullerenes

Fullerenes are symmetrical closed convex graphite shells, consisting of 12 pentagons and various numbers of hexagons, for which $K > 0$. Some are tessellations of the icosahedron. They have been illustrated recently by Smalley & Curl (1991) and by many others so that it is not necessary to discuss them further here. Being topologically equivalent to the sphere they have genus 0 and faces+vertices = edges+2. Further, $N_5 - N_7 - 2N_8 = 12$.

9. Periodic minimal surfaces

Minimal surfaces are surfaces with $H = 0$ so that $K_1 = -K_2$ and $K \leqslant 0$. They are thus saddle-shaped (anti-clastic) everywhere except at certain 'flat points' which are higher order saddles (for example the 'monkey-saddle' which has symmetry $\bar{3}$). A surface may be minimal either because, as for a soap-film spanning a non-planar loop of wire, it minimizes its energy by having a minimum of area or, for a membrane surface made of lipid molecules, because it minimizes the splay energy. The mathematical condition for a surface to have zero mean curvature is that the divergence of its unit normal should be zero. The splay energy is the integral of H^2 over the area. Surfaces for which the integral of H^2 is a minimum are called Willmore surfaces.

It is not possible to construct an infinite surface with constant negative gaussian curvature. Such a surface with a constant, imaginary radius of curvature defines the hyperbolic plane H^2 (the surface of a sphere being designated as S^2).

However, H. A. Schwarz found before 1865 that patches of varying negative gaussian curvature and constant $H = 0$ could be smoothly joined to give an infinite triply periodic surface of zero mean curvature. About five different types were found by Schwarz and Neovius, but now about 50 more have been described (Schoen 1970; Fischer & Koch, 1989*a–e*).

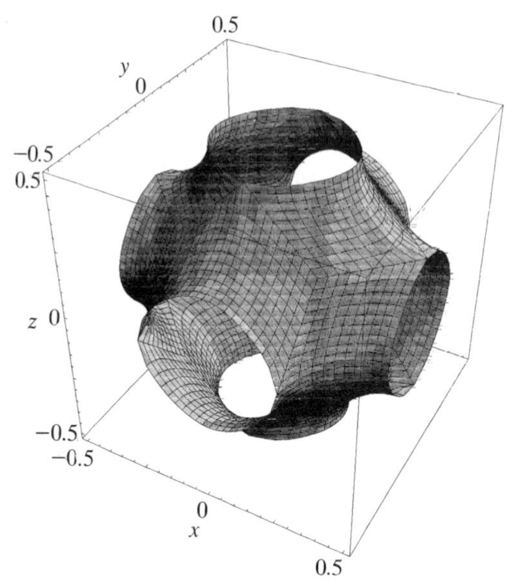

Figure 1. The *P*-surface, found by H. A. Schwarz. If the two sides are the same, the space group is Im$\bar{3}$m: if different, then Pm$\bar{3}$m. It corresponds closely to the zero equipotential surface in CsCl.

One case is of a tetrahedral frame of four rods over which a soap-film may be placed to give a saddle-shape. Such pieces, each of eight asymmetric units, can be joined smoothly with diad axes along the rods to give a continuous surface, the *D*-surface, which is triply periodic and which divides space into two congruent regions. This surface can be imagined in various ways; one way is to take the structure of diamond and to inflate the bonds into tubes until the space between the tubes is congruent with the space inside the tubes. It thus has the symmetry of the cuprite ('double diamond type') structure of space group Pn$\bar{3}$m. However, the *P*-surface shown in figure 1 is the simplest to apprehend and we will preferentially illustrate its variants.

Two classes of periodic minimal surfaces may be distinguished. By examining all the asymmetric regions in the 230 crystallographic space groups, Fischer & Koch (1989*a–e*) enumerated completely all the 'balanced surfaces', where, as in the D-surface, the two sub-spaces are congruent. There is also an indefinitely large class of unbalanced surfaces where the two subspaces are not congruent. On minimal surfaces every point has zero mean curvature and non-positive gaussian curvature but at certain points, flat points or umbilics, both principal curvature are zero and the surface is completely determined once the orientation of its normals at the flat points is given. Surfaces may again be divided into those for which the flat points are symmetrically equivalent and those for which this is not so. Self-intersecting surfaces, which are at present not of physical importance, have been neglected.

Exact procedures for determining the shapes of the periodic minimal surfaces are available and have been described elsewhere (Nitsche 1975, 1989; Terrones 1992; Fogden & Hyde 1992*a, b*). They involve representing each point on the surface by a point in the complex plane. The coordinates x, y, z of a point in the surface are related to this complex number ω by stereographic projection from the Gauss sphere, on which the normal directions to each point on the surface are marked, and thence to the actual points in space by the Weierstrass integrals which can be computed being, in the simpler cases, related to elliptic integrals. Each surface has a characteristic Weierstrass function $R(\omega)$ (Fogden & Hyde 1992*a*, *b*) although in some cases this is excessively complicated but, for example, Schwarz' D-surface has $R(\omega) = (1-14\omega^4+\omega^8)^{-\frac{1}{2}}$.

Certain surfaces are related and, for the D, P and G surfaces, the coordinates of one are obtained from those of another by multiplying the complex number ω by $\exp(\mathrm{i}\theta)$ where θ is the Bonnet angle. The catenoid has the simplest Weierstrass function $R(\omega) = 1/\omega$ and the helicoid has the same function but multiplied by $-\mathrm{i}$ so that one surface can be bent from one to the other without local distortion (change of metric). If the infinite surfaces are continuous, then they would have to pass through each other, but if we consider only discrete atoms lying in conceptual surfaces, then this constraint does not apply. As an equivalent, this Bonnet transformation can be seen as the twisting of a hexagonal patch (having three vertices turned up and three down: the 'monkey saddle'). Such patches in a continuous surface are separated, twisted and then joined together again differently.

Hyde & Andersson (1986) have explained martensitic transformation in these terms and provided some evidence for it. Even if this is eventually not substantiated, the idea is certainly *molto ben trovato* and merits development. It is a remarkable materialization of a hidden phase factor.

Surfaces may found by finite element analysis methods where the curvature of each element of surface is brought iteratively to the correct value. More general energy functions can be imposed in this way. Exact minimal surfaces are merely particular idealizations and their value lies in their being two-dimensional manifolds which have metrics different from that of the euclidean manifold of the plane.

Surfaces can also be represented as the zero equipotentials between positive and negative ions arranged as in a crystal. Periodic minimal surfaces are often close to the zero equipotentials in real crystal structures.

Mathematical approximations to the periodic minimal surfaces can be constructed from terms which are each the result of adding symmetry-related sinusoidal density waves for the appropriate symmetry group, and then taking the nodal surface; the boundary between regions of positive and of negative density. The waves that correspond to a face-centred figure in real space are the body-centred terms in reciprocal space, namely:

$$\begin{aligned} &C_0+C_{110}(3-\Sigma\cos 2\pi x\cos 2\pi y)+C_{200}(1-\Sigma\cos(2\pi 2x)) \\ &\quad +C_{211}(6-\Sigma\cos(2\pi 2x)\cos 2\pi y\cos 2\pi z)+C_{220}(3-\Sigma\cos(2\pi 2x)\cos(2\pi 2y)) \\ &\quad +C_{310}(6-\cos(2\pi 3x)\cos(2\pi y))+C_{222}(1-\cos(2\pi 2x)\cos(2\pi 2y)\cos(2\pi 2z))+\dots. \end{aligned}$$

For a primitive cubic lattice the first approximant is:

$$\cos(2\pi x)+\cos(2\pi y)+\cos(2\pi z) = 0.$$

This later expression is very convenient for generating a close approximation to the P-surface. Townsend *et al.* (1992) have used it and other such expressions as manifolds on which to construct irregular sphere packings which can represent (after removing the spheres at the centres of rings of 5, 6 or 7 spheres) the graphite network.

Random surfaces can be constructed by adding sine waves of random amplitude, direction and phase, but all with the same wavelength, and then taking the nodal surface where the value of the function is zero. We have applied stereological methods to estimate statistically the area and curvature of such surfaces which are

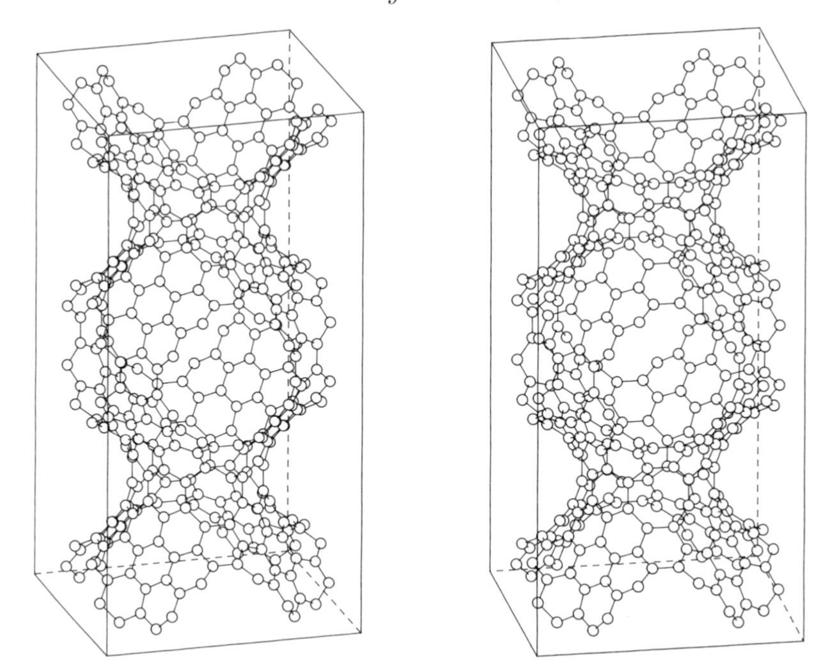

Figure 2. A possible graphite structure based on the P-surface, where the network of hexagons is curved by the introduction of octagons (stereo-pair of two unit cells).

available for decoration by graphite nets. The area per unit volume is close to that of regular periodic minimal surfaces with the same characteristic length.

10. Graphite structures with negative gaussian curvature

By decoration of these various infinite two-dimensional manifolds (just as the sphere has been decorated with closed networks) several related structures have been proposed for graphite nets. These are mostly based on the P, D and G surfaces (the first two due to Schwarz (1890) and the last, the gyroid, discovered by Schoen (1970). However, many other surfaces (perhaps 50) are available for consideration. Some fit naturally with hexagonal sheets and others with sheets of square or lower symmetry. In general, the P, D and G surfaces are the least curved from planarity. Surfaces parallel to the surfaces of zero mean curvature have lower symmetry than those with $H = 0$. When decorated with graphite nets the symmetry may be further lowered to that of a sub-group of the symmetry group of the surface itself.

(*a*) *Mackay and Terrones*

We have successfully built models of the P, D, G and H surfaces (by computer graphics and physically by using three-way joints and connector tubes to represent the graphite net) (Mackay & Terrones 1991) (figures 2 and 3). Triangular patches of various sizes of the graphite network with angles 90° 30° and 45° can be joined so that the 45° vertices combined to give rings of eight carbon atoms. In the actual structures these sit appropriately on the saddle points of highest local gaussian curvature. Given the basic patch, the asymmetric unit of pattern, this can be twisted to give either the P, the D or the G surfaces.

In Mackay & Terrones (1991) 144 points of the type 8.6^2 and 48 of the type 6^3 make

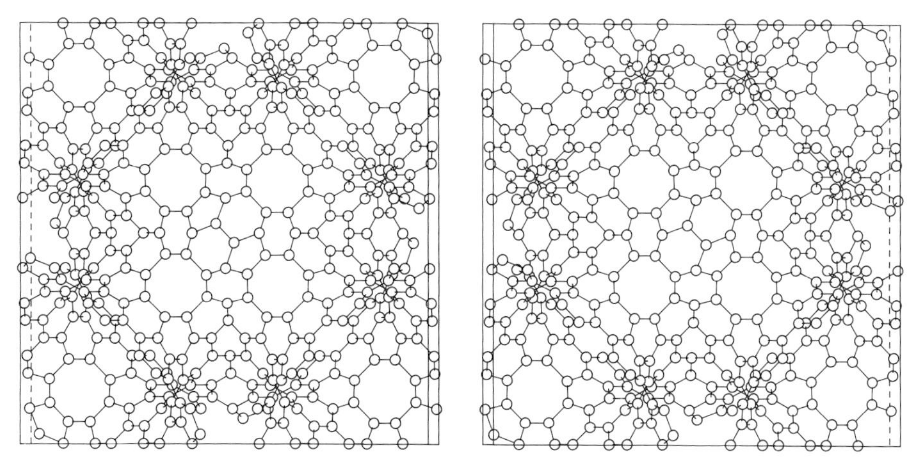

Figure 3. A decoration of the *I-WP* surface (stereo-pair of half of the Im$\bar{3}$m unit cell.)

up the 192 vertices per cell. In each case any number of 6^3 points may be added or removed without changing the vertex excess. We might also consider points of the type $8^2.5$ of which 80 per cell of genus 3 would be needed plus any number of 6^3 points but this would be unrealistic for graphite.

The polyhedron corresponding to the Neovius surface has the same arrangement of points as that for the infinite semi-regular polyhedral surface 6.4^3 discussed above, but the spaces between the points are differently filled with polygons so that each of the 48 points per cubic cell has the configuration of 8.4.8.6 and this leads to a surface of genus of 9. This surface has two kinds of flat points and is thus not 'regular' (Mackay & Terrones 1991). 12 tubes in the [110] directions connect cavities.

The *I-WP* surface (in Schoen's (1970) idiosyncratic notation) is an unbalanced cubic surface where eight tubes in the [111] directions connect cavities. It is close in shape to the Fermi surface of the FCC metals Cu, Ag, Au. In our decoration (figure 3) (with rings of 8, 6 and 5 atoms) the symmetry is reduced from that of the surface, which is Im$\bar{3}$m, by the replacement of 4-fold axes by 2-folds. There are 228 hexagons, 48 octagons and 24 pentagons per cubic cell. The pentagons can be introduced in several ways.

(*b*) *Lenosky, Gonze, Teter and Elser*

Lenosky *et al.* (1992) have a somewhat different combination where 216 vertices of the type 7.6^2 and 48 of the type 6^3 add to give the vertex excess of 8π (figure 4). The first two of their surfaces are parallel to the *P* and *D* periodic minimal surfaces and divide space into two unequal regions (and are thus less symmetrical than the periodic minimal surfaces). It would be possible to arrange two identical parallel surfaces to be 3.4 Å apart by increasing the size of the asymmetric patch appropriately. The atomic positions have been refined and physical properties of the structures have been calculated. The asymmetric patches contain heptagons rather than octagons to introduce the negative gaussian curvature and such structures are found, by calculation, to be energetically more favourable than spherical buckminsterfullerenes.

Lenosky, Gonze, Teter and Elser propose the name schwarzites, in memory of H. A. Schwarz, for this category of graphites with non-positive curvatures. Since the name seems vacant as a mineral name, we commend the proposal.

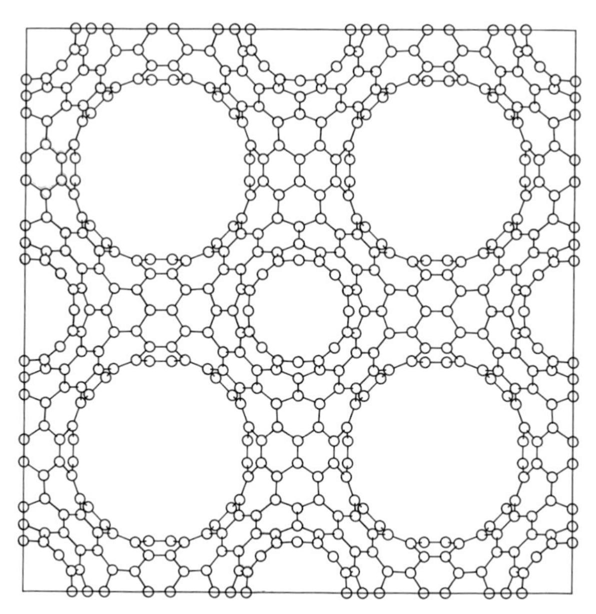

Figure 4. Four unit cells of a possible graphite structure where the introduction of heptagons enables the graphite net to be mapped on to a surface parallel to the *P*-surface (Lenosky *et al.*).

We are much indebted to Lenosky for calculating, on the same basis, energies of our structures, those of Vanderbilt and Tersoff and of his own (table 1), so that their stabilities can be compared with that of C_{60}. It will be seen that there are several structures with energies less than that of C_{60}.

Townsend *et al.* (1992) later produced several similar structures where the network lies in the minimal surface. In particular, they have also constructed random surfaces and have made a good *prima facie* case that seaweed-like amorphous graphite actually exists. Experimental electron diffraction scattering corresponds well with that calculated from minimal surface structures.

(*c*) *Vanderbilt and Tersoff*

To make a surface of genus 3 (as for the *P*, *D*, and *G* surfaces) the excess vertex sum for the vertices within the unit cell should be 8π. This is given by 56 vertices of the type 7^3 and is the surface from which Vanderbilt & Tersoff (1992) began to develop their model. If this tessellation is now truncated, that is, a hexagon is placed at each vertex, the tessellation becomes 7.6^2 (with all the vertices still equivalent to each other). To make up 8π 168 vertices are needed. This number gives a more relaxed structure than that using heptagons alone. The process is exactly parallel to the truncation of the regular dodecahedron, with vertices 5^3, to give the truncated icosahedron 5.6^2 of C_{60}.

The surface is one parallel to the *D*-surface, so that the two sub-spaces are not equivalent. A tetrahedral joint is built out of 84 atoms in hexagons and heptagons so that each point is a member of two hexagons and one heptagon ($6^2.7$). There are thus 2×84 atoms per primitive unit cell with space group Fd3 and $8 \times 84 = 672$ per cubic unit cell with $a = 21.8$ Å (assuming graphite-type bonds).

The density is expected to be 1.29 g cm^{-3} and the authors calculate the energy of formation to be 0.11 eV atom^{-1} as compared with 0.67 eV atom^{-1} for C_{60}.

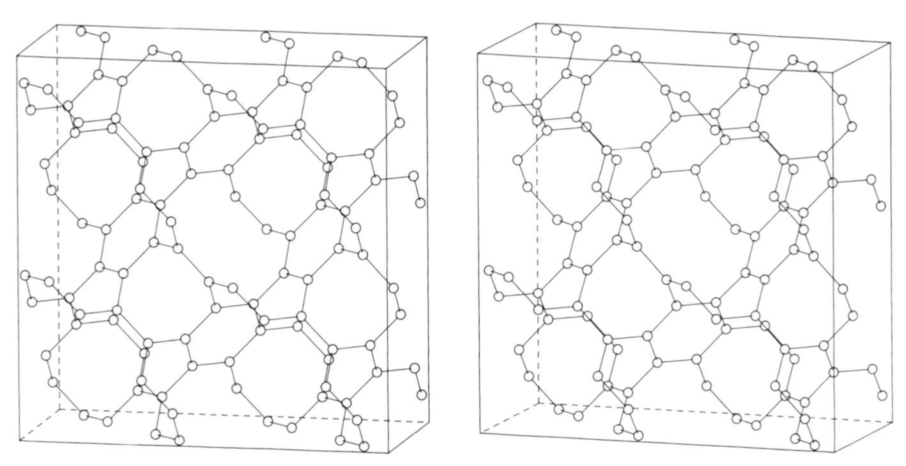

Figure 5. The simplest decorated *D*-surface, 'polybenzene', found by O'Keefe *et al.* (the stereo-pair shows 4 unit cells of space-group Pn$\bar{3}$m, each containing 24 atoms).

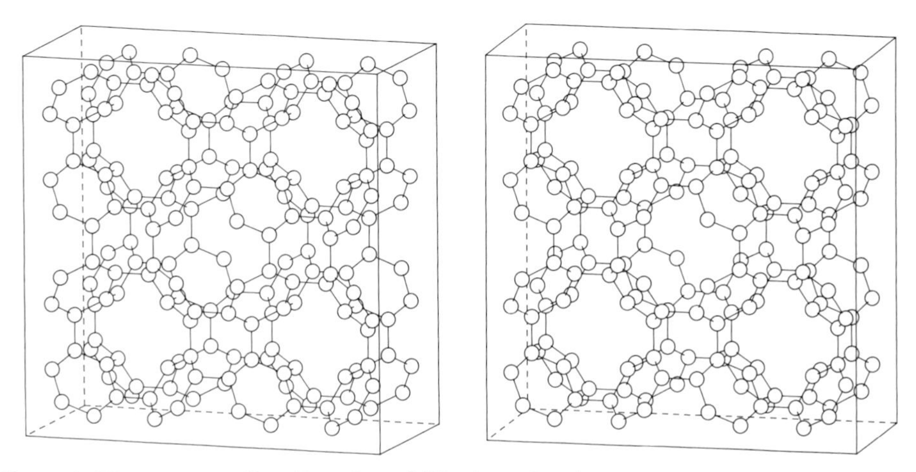

Figure 6. The corresponding *P*-surface (O'Keefe *et al.*) (the stereo-pair shows 4 unit cells of space-group Im$\bar{3}$m, each with 24 atoms).

(*d*) *O'Keefe, Adams and Sankey*

O'Keefe *et al.* (1992) have reduced the asymmetric patch (in the minimal surface) to the minimum and have examined the corresponding *P* and *D* surfaces (presumably the related *G* surface could also exist). These are significant in that all the atoms can now be equivalent and the vertices are 6.8^2. The 6.8^2 *D* structure is particularly interesting and had been reported before by Gibson, Holohan and Riley and by Wells (1977) and can be referred to as 'polybenzene'. It has a considerably lower energy than that of C_{60}, but that for the 6.8^2 *P* form is a trifle higher. The densities are much greater than for the other proposed surfaces. Figure 5 shows the *P*-surface and figure 6 the corresponding *D*-surface.

It is important to realize that not all three-connected nets in three-dimensions lie on surfaces. A great variety of both nets and surfaces has been described, somewhat cryptically, by Wells (1977).

Altogether, the variety of possible structures and their low energies compared with that of C_{60}, together with the measurement by Elser *et al.* of an electron scattering

Table 1. *Comparison of various hypothetical structures*

	structure	ρ	ΔE	$a/\text{Å}$	N	spacegroup
1	P216 bal.	1.11	0.17	15.7	108 × 2	Ia3 (?)
2	D216 bal.	1.16	0.16	15.50 (24.6)	216 (× 4)	Pn3 (Fd3 ?)
3	random 1248	1.26	0.23	—	—	—
4	G216 bal.	1.18	0.17	—	216	$Ia\bar{3}d$
5	D7 par.	1.15	0.18	24.7	216 × 4	$Fd\bar{3}m$
6	P7 par.	1.02	0.20	16.2	216	$Pm\bar{3}m$
7	P8 bal.	1.16	0.19	14.9	192	$Im\bar{3}m$
8	D8 bal.	1.10	—	15.16 (24.09)	192 (× 4)	$Pn\bar{3}m$
9	G8 bal.	1.12	—	18.94	192 × 2	$Ia\bar{3}d$
10	*I-WP*	1.06	—	24.09	744	$I\bar{4}3m$
11	D7 par.	1.28	0.22	21.8	168 × 4	Fd3
12	6.8^2P bal.	2.04	0.488	7.770	48	$Im\bar{3}m$
13	6.8^2D bal.	2.19	0.208	6.033	24	$Pn\bar{3}m$
	FCC C_{60}	1.71	0.42	14.12	60 × 4	—
	diamond	3.52	0.02	3.5595	2 × 4	$Fd\bar{3}m$
	graphite	2.28	0	$a = 2.460$		

ρ is the calculated density in g cm^{-3}.
ΔE is the total energy relative to graphite in eV $atom^{-1}$.
a is the cubic unit cell size.
N is the number of C atoms per unit cell (the FCC cell contains 4 primitive cells and the BCC cell 2).
bal. indicates a balanced surface, where the two subspaces are congruent; par. marks a less symmetrical surface parallel to this.
1, 2, 3, 4, 5, 6 are from Lenosky *et al.* (1992) and Townsend *et al.* (1992). They contain heptagonal rings.
3 is a random covering of the D-surface type with $a = 42.9$ Å with 1248 atoms. In this unit there are 38 pentagons, 394 hexagons, 155 heptagons, 12 octagons and 1 nonagon.
7, 8, 9, 10 are due to Mackay & Terrones (1991) for C–C taken as 1.42 Å. They contain octagonal rings. (The *I-WP* surface contains also pentagons: it is not balanced.)
The value 14.9 Å was obtained by Lenosky after refinement from our model.
11 is due to Vanderbilt & Tersoff (1992). It uses heptagonal rings.
12 and 13 are due to O'Keefe *et al.* (1992) and contain octagonal rings. Their positional parameters for the P structure are $x = 0.3103$ and $z = 0.0867$ and for the D structure, $x = 0.3364$. 13 is 'polybenzene'.

pattern, strongly indicate the possible existence of graphite surfaces of negative gaussian curvature, rather like seaweed, which, only under rather special circumstances, such as production by charring of a ordered liquid crystal, might be three-dimensionally periodic.

Table 1 collects data from various authors. Note that the space group of the symmetrical D-surface (symmetry of cuprite or 'double diamond', containing two tetrahedral joints in opposite orientations) (in which atoms may be embedded less symmetrically (as in 2) is $Pn\bar{3}m$. If the structure is made less symmetrical (the diamond structure) by taking the parallel surface, then the contents of the unit cell are multiplied by 4 and the space group becomes $Fd\bar{3}m$ (containing 4 primitive rhombohedral cells in each of which there are two tetrahedral joints in opposite orientations). The cubic cell side is multiplied by $2^{\frac{2}{3}}$. If atoms are embedded less symmetrically, then the space group becomes a subgroup of the space group of the surface.

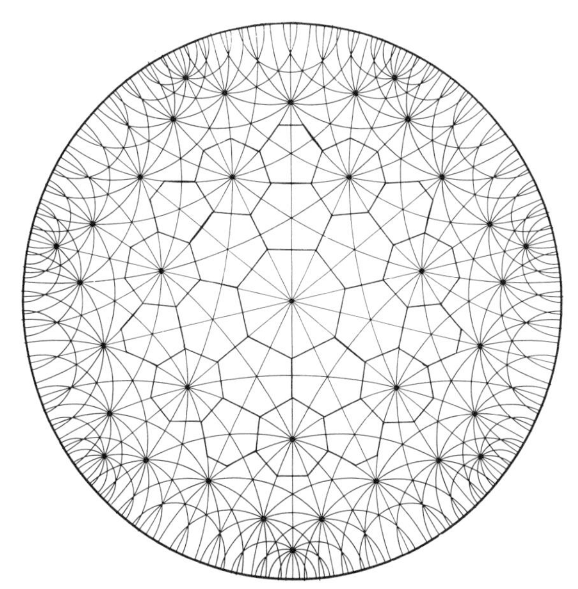

Figure 7. A tessellation of the hyperbolic plane H^2 by regular heptagons. The 7.6^2 tessellation (thicker lines) is obtained by truncation, a hexagon replacing each three-fold vertex. Polygons are connected as they are locally in Vanderbilt and Tersoff's surface.

11. The hyperbolic plane, H^2

The hyperbolic plane (called H^2, the surface of the sphere being S^2), has a constant negative gaussian curvature. Regular and irregular tessellations of this surface by polygons can represent topologically the local connectivities of real space polygonal and continuous surfaces which have everywhere non-positive curvatures. Just as a closed spherical surface can be covered by a planar pattern by regular disclinations (where a sector is cut out and the cut edges are joined) so a hyperbolic surface can be made by regularly inserting a sector. For example a regular net with hexagonal symmetry can have sectors inserted to make sevenfold symmetry.

In particular, the regular tessellation of H^2 by heptagons (figure 7) (Mackay 1986) can be truncated to give the tessellation 7.6^2 which corresponds to Vanderbilt and Tersoff's surface and is the least curved of the regular tessellations. Further truncations would increase the ratio of hexagons to heptagons, but all vertices would then be no longer equivalent to each other. When we attempt to execute the H^2 tilings in real space the curvature causes the surface to fold like sea-weed (as *fucus letuca*) since more area is produced within a given radius than is appropriate for a planar tiling (Thurston & Weeks 1984). After a certain number of units have been added, the surface can be closed on itself in various ways, periodically or irregularly. Just as in producing cylinders, one vector in the plane tessellation is identified as corresponding to circuits of the cylinder so, in the hyperbolic plane, cycles can be found in several directions.

12. Transformations and kinematics

There are many topics for further investigation.

We have referred above to the promise of the Bonnet transformation in describing actual martensitic transformations. As a generalization of the catenoid/helicoid transformation at constant gaussian curvature (here zero), it has long been known that any surface of revolution can be bent into a screw surface. Other transformations have been developed by Gackstatter and by Kenmotsu and by Terrones (1992) which, although requiring some distortion of the surface, produce transformations, such as the turning of a slit sphere inside out, which may be of relevance, if not for graphite, then for the phenomena of vesicles, such as gastrulation. Computer graphics have much facilitated these applications. The mechanism of the coalition of two C_{60} or C_{70} particles (Yeretzian *et al.* 1992) may require some such visualization and may connect with the possible processes of formation of minimal surfaces.

The trajectories of rays, representing particles or waves, inside one of the periodic minimal surfaces are of great relevance in considering the behaviour of, for example, photons, reflected inside such a labyrinth. By following these trajectories one can observe what regions, if any, of the space are preferentially visited. After kinematics, dynamics must be investigated. The surfaces discussed above have many applications in fields other than that of graphite.

References

Ajayan, P. M. & Iijima, S. 1992 Smallest carbon nanotube. *Nature, Lond.* **358**, 23.

Bernal, J. D. 1933 Discussion on liquid crystals. *Trans. Faraday Soc.*, p. 1081.

Erickson, R. O. 1973 Tubular packing of spheres in biological fine structures. *Science, Wash.* **181**, 705–716.

Fischer, W. & Koch, E. 1989*a* New surface patches of minimal balance surfaces. I. Branched catenoids. *Acta crystallogr.* A **45**, 166–169.

Fischer, W. & Koch, E. 1989*b* II. Multiple catenoids. *Acta crystallogr.* A **45**, 169–174.

Fischer, W. & Koch, E. 1989*c* III. Infinite strips. *Acta crystallogr.* A **45**, 485–490.

Fischer, W. & Koch, E. 1989*d* IV. Catenoids with spout-like attachments. *Acta crystallogr.* A **45**, 558–563.

Fischer, W. & Koch, E. 1989*e* Genera of minimal balance surfaces. *Acta crystallogr.* A **45**, 726–732.

Fogden, A. & Hyde, S. T. 1992*a* Parametrisation of triply periodic minimal surfaces. I. *Acta crystallogr.* A **48**, 442–451.

Fogden, A. & Hyde, S. T. 1992*b* II. *Acta crystallogr.* A **48**, 575–591.

Fourcade, B., Mutz, M. & Bensimon, D. 1992 Experimental and theoretical study of toroidal vesicles. *Phys. Rev. Lett.* **68**, 2551–2554.

Gennes, P. G. de 1992 Soft matter. *Science, Wash.* **256**, 495–497.

Hyde, S. T. & Andersson, S. 1986 The martensite transition and differential geometry. *Z. Kristallogr.* **174**, 225–236.

Iijima, S. 1991 Helical microtubules of graphitic carbon. *Nature, Lond.* **354**, 56–58.

Iijima, S., Ichihashi, T. & Ando, Y. 1992 Pentagons, heptagons and negative curvature in graphite microtubule growth. *Nature, Lond.* **356**, 776–778.

Klug, A., Crick, F. H. C. & Wyckoff, H. W. 1958 Diffraction by helical structures. *Acta crystallogr.* **11**, 199–213.

Lenosky, T., Gonze, X., Teter, M. & Elser, V. 1992 Energetics of negatively curved graphitic carbon. *Nature, Lond.* **355**, 333–335.

Mackay, A. L. 1985 Periodic minimal surfaces. *Nature, Lond.* **314**, 604–606.

Mackay, A. L. 1986 Two-dimensional space groups with sevenfold symmetry. *Acta crystallogr.* A **42**, 55–56.

Mackay, A. L. & Terrones, H. 1991 Diamond from graphite. *Nature, Lond.* **352**, 762.

Nitsche, J. C. C. 1975 *Vorlesungen über Minimalflächen.* Berlin: Springer-Verlag.

Nitsche, J. C. C. 1989 *Lectures on minimal surfaces*, vol. 1. Cambridge University Press.

O'Keefe, M., Adams, G. B. & Sankey, O. F. 1992 Predicted new low energy forms of carbon. *Phys. Rev. Lett.* **68**, 2325–2328.

Schoen, A. H. 1970 Infinite periodic minimal surfaces without self-intersections. NASA Technical Note D-5541.

Schwarz, H. A. 1890 *Gesammelte Mathematische Abhandlungen*, 2 vols. Berlin: Springer.

Smalley, R. E. & Curl, R. F. 1991 The fullerenes. *Scient. Amer.* **265**, 32–37.

Terrones, H. 1992 Mathematical surfaces and invariants in the study of atomic structures, Ph.D. thesis, University of London, U.K.

Thompson, D'A. W. 1925 On the thirteen semi-regular solids of Archimedes, and on their development by the transformation of certain plane configurations. *Proc. R. Soc. Lond.* A **107**, 181–188.

Thurston, W. P. & Weeks, J. R. 1984 Three-dimensional manifolds. *Scient. Amer.* **251**, 94–106.

Tibbetts, G. G., Devour, M. G. & Rodda, E. J. 1987 An adsorption-diffusion isotherm and its application to the growth of carbon filaments on iron catalyst particles. *Carbon*. **25**, No. 3, 367–375.

Townsend, S. J., Lenosky, T. J., Muller, D. A., Nichols, C. S. & Elser, V. 1992 Negatively curved graphite sheet model of amorphous carbon. *Phys. Rev. Lett.* **69**, 921–924.

Vanderbilt, D. & Tersoff, I. 1992 Negative-curvature fullerene analog of C_{60}. *Phys. Rev. Let.* **68**, 511–513.

Wells, A. F. 1977 *Three-dimensional nets and polyhedra*. New York: Wiley.

Whittaker, E. J. W. 1957 *Acta crystallogr.* **10**, 149 and earlier papers.

Yada, K. 1971 Study of microstructure of chrysotile asbestos by high resolution electron microscopy. *Acta crystallogr.* A **27**, 659–664.

Yeretzian, C., Hansen, K., Diederich, F. & Whetten, R. L. 1992 Coalescence reactions of fullerenes. *Nature, Lond.* **359**, 44–47.

Discussion

P. W. Fowler (*University of Exeter, U.K.*). Please comment on finite analogues of your structures? We have discussed previously the idea of making a 'Russian doll' fullerene in which one fullerene is connected by tunnels to an outer shell. The matrix seems to comply with this.

A. L. Mackay. We have carried out model building experiments and you can indeed close off tubes with domes, more or less at will.

P. W. Fowler. Have you had the same experience as us? If you leave a model around it bursts spontaneously. I think the steric strain therein is very great.

A. L. Mackay. Our open models are surprisingly strain-free, given that you have equal bond lengths and 120° connections.

P. W. Fowler. If you make a model with icosahedral symmetry I believe you will find that you will be left with plastic fragments.

A. L. Mackay. Yes, if you use the same materials to build C_{60} it is clearly strained. But I should say that the definition of a minimal surface is that it has zero mean curvature; i.e. you can say that either the divergence of the normal is zero, or that it has zero splay. Thus if you take the p-orbitals in one direction they are spread out, whereas in another they are compressed or folded in. So the whole evens out. Minimal surfaces have, characteristically, zero splay energy. This is an argument in its favour, whereas spheres are splayed in both directions.

Fullerenes as an example of basic research in industry

By E. Wasserman

Central Research and Development, The Du Pont Company, Wilmington, Delaware 19880-0328, U.S.A.

Some recent trends in industrial basic research are considered. They are driven by a highly competitive global marketplace. The discussion focusses on two recent major scientific advances which have been pursued in industry: high-temperature superconductivity and C_{60} with its chemical family. Fullerenes appear to be appropriate candidates for basic research in industry.

This paper might be subtitled 'Research on a Globe, Chemistry on a Sphere', referring to two very different aspects of the chemical community involved with fullerenes. Research worldwide is undergoing major transformations. Although noticeable in academic and government laboratories, the changes are larger in the industrial sector, especially in companies that have been major centres of basic research for decades. Here we shall consider some of the more prominent developments in industry.

The second aspect is the extraordinary chemistry of C_{60} and the rapidly increasing family of carbon structures. The chemistry will be that developed at Du Pont; the work of Dr Fagan, Dr Krusic and Dr Tebbe with others. Taylor's contribution in this collection provides an extended consideration of some of the topics and will not be repeated here. Rather we focus on a few features of chemistry on a sphere as opposed to the planar or linear compounds with which we are more familiar.

The bridge between these two parts, and a conclusion, is that basic research on fullerenes may well be an appropriate activity for a modern company dealing with chemicals or materials.

Changes in research philosophy and practice have occurred throughout technologically based industry in recent years. As companies position themselves for a competitive, fast-moving marketplace the fundamental importance of research and technology is rarely questioned. But there is continual discussion as to what research should be done, how the research should be carried out, what are the proper roles of industrial, governmental and academic laboratories, and how the results should be transferred and developed for the commercial realm.

These questions are not new. At Du Pont similar ones were being asked at the beginning of the century and have reappeared every ten to twenty years. They are with us now. With changing circumstances, both external and internal to the corporation, the answers are also changing.

Some of the more frequent comments throughout industry are the following.

1. *Build on company strengths:* Now we are less likely to develop a new business from research as man-made polymers and transistors were created in decades past. Pharmaceutical companies continue to produce novel products from their base in

Phil. Trans. R. Soc. Lond. A (1993) **343**, 129–132

Printed in Great Britain

molecular biology and chemical synthesis. But they focus on areas closely related to those they know well rather than branching into fundamentally new fields.

2. *Faster progress in research.* The increasing cost of research in industry often outpaces consumer-oriented indices of inflation. There is a continuing demand for more rapid developments. This increased rate requires that old habits be changed, organizations more tightly structured, collaborations more effective and movement to market faster than previously.

3. *Shorter programmes.* While related to 2, this requirement highlights the penalty of greater length. We seek 3–5 year programmes rather than 10–15 years. The extended programme can become obsolete before completion. Personnel movement between companies together with the information available in patents and publications puts the competitive success of a longer programme at risk. Sometimes there is an advantage being the second to enter the race, avoiding much of the time-consuming problem-solving required of the pioneer. The increased pace is not unreasonable as new tools, often analytical, allow completion of a programme in less time.

It is not appropriate here to comment on these new rules except that they are consistent with the competitive global marketplace which must be considered in planning industrial activities. My guess is that by the end of the decade we shall be seeing a return to longer term fundamental research in industry but in areas different from those that are now popular. But that is another story.

For now I should like to consider two fields in which recent breakthroughs in fundamental science have led to major efforts in industry and examine some critical differences between them. One, of course, is the fullerenes and we shall return to that below.

The other is high temperature superconductivity which began with the report of Bednorz & Müller (1986). Within a year the upper limit of T_c went from 23 K to 125 K an extraordinary accomplishment given that sixty years had been required for T_c to rise from 4 K to 23 K. Within weeks of the first discoveries popular accounts discussed the wonders that might be possible in energy saving and production together with major changes in society that could be envisioned. While important scientific advances continue to appear, the commercially significant applications are still many years away. The gaps between possibilities and accomplishments has led some to question our ability to judge the likely benefits of basic research for industry. Of course, those of us who have been involved with such research we are well aware of the surprising turns, both positive and negative, which can occur. We emphasize the positive as we must; we are among the best qualified to see possible advantages. But the qualifiers initially included with the conjectures are often lost. The apparent credibility gap can lead higher management to wonder about the level of support appropriate for more fundamental research programmes.

Superconductivity highlighted another complication for basic research with possible industrial applications; the outstanding quality of laboratories around the globe and the consequent high level of competition. With many organizations starting with similar levels of expertise the probability that the efforts of one will lead to a commercially successful project is small. The sheer numbers involved reduce the chance that a given company will win the race. Also, proprietary rights can be confused as different companies and universities contribute to the development of a single product.

In earlier times, in some of the classic cases of fundamental research leading to successful new businesses, technical competition was much less. For polymer research at Du Pont in the late 1920s and 1930s industrial competition was centred at I. G. Farben in Germany. The programme under Wallace Carouthers at Du Pont was able to set its separate direction producing polyamides and neoprene for the marketplace while developing the fundamental science which established polymer chemistry as a rigorous intellectual discipline.

Similarly, Bell Laboratories pursued the transistor before and after World War II when solid state physics was not heavily populated. Bell could set its own pace in reasonable confidence that they would have a strong proprietary position. There were also features unique to the Bell System that are unlikely to be duplicated in today's competitive marketplace. Bell was a regulated monopoly with its profits guaranteed by government regulation. In addition, the System's internal needs provided a large market for any new devices.

Another feature is associated with the high level of activity in basic research throughout the industrial world, namely considerable duplication and consequent inefficient use of resources in a suddenly fashionable area. A given experiment may be carried out simultaneously in many laboratories with only a small probability of a worthwhile return from the individual effort.

One example, more extreme than most hopefully, occurred at a meeting of the Material Research Society a year after the initial explosion of interest in superconductivity. During a symposium mention was made of a recently published experiment reporting an increase in T_c by treatment of the solid superconductor with a gas. The chairman asked how many members of the audience had repeated this experiment. Fifty hands were counted out of a total of 450. It is likely that others, not in the audience, had also tried. Almost all of this effort is an unproductive use of the resources alloted to the field. Little appeared in the literature as negative results with such complex solids are rarely definitive and publishable. But remedies are not readily available.

In many countries the freedom of the individual investigators to decide the details of their research programme is well protected. Control of day-by-day activities is inappropriate. Nevertheless, in cases such as the above, a preferable procedure would be for two or three selected laboratories to attempt confirmation of the claim. They would then communicate the results to others. Again this massive duplication may be regarded as a corollary of the large number of well-staffed and well-supported research laboratories around the globe.

Another feature of the superconductivity effort has been that most has been devoted to detailed studies of existing materials rather than wide-ranging searches for new compositions. Substantially enhanced properties are necessary if high temperature superconductors are to have a major impact in the commercial sector. The assumption has been that detailed understanding will lead to improved superconductors. Usually, this has not occurred. The complexities of these underdetermined solid state systems restrict the ability to design improved structures.

The McCall committee on high-temperature superconductivity reported that some two years after the breakthrough in the field roughly US$1000 M worldwide was being spent on research and development while the total market for superconductivity of all types was but US$250 M. Of course the research and development expenditures are in search of new capabilities and markets. While larger markets are possible the area is likely to be unprofitable for years to come.

High-temperature superconductivity initially attracted more than a thousand participants globally, but after one and a half to two years many left the field. Much of that migration occurred because the near-term commercial opportunities appeared limited. While the science remained fascinating and was an extraordinary advance in superconductivity there were fewer major discoveries. The theory of the phenomenon and the related normal state of these unusual materials continues to attract the attention of some of the best solid-state theorists. We can expect major additional scientific developments in coming years.

The shrinking effort in industry has led to smaller, more focused programmes on a scale appropriate to potential markets. The activities have comparatively modest goals. A number of products should appear in the near-term, primarily in electronics with fewer in energy-related fields and transportation.

In contrast, fullerenes after two years of effort are continuing to attract more practitioners to the field. Related areas of research are appearing such as giant and nested fullerenes as well as buckytubes. Such growth and branching is a sign of a developing discipline.

One of the great attractions of fullerenes is the ability to apply to C_{60} much of the olefin and aromatic chemistry already developed with traditional organic compounds. This activity is still in its early stages. One of the earliest accomplishments was the use of organometallic chemistry by Paul Fagan to demonstrate that C_{60} could react as an electron-poor olefin similar to tetracyanoethylene.

Duplication in fullerene chemistry is substantially less than that found with high-temperature superconductors. Several high temperature superconductors are easily made. The science involved in purifying, characterizing and reacting fullerenes can be demanding. At times C_{60} can be as recalcitrant as a piece of coal. It may be the purest form of carbon before reaction but the products are often impure derivatives difficult to separate. The well-defined compounds that have been obtained often precipitated from solution under equilibrium conditions as with Fagan's platinum complexes and Tebbe's $C_{60}Br_{24}$. These can easily be redissolved and transformed.

While the ability to relate the chemistry of C_{60} to that of other organic molecules has been beneficial for many studies, the novel reactivity of the spherical form is of particular interest. The quantitative thermal decomposition of $C_{60}Br_{24}$ to C_{60} and Br_2 is one example with little precedent in traditional reactions. Another is the observation by Paul Krusic that up to thirty-four methyl radicals can be added to the fullerene. This holistic behaviour is promising for other novel chemistry.

As a field of basic research in industry fullerenes are attractive as there are a number of possible applications. Some of these could provide new business opportunities. In the short term extensions of existing businesses and variations of existing products are likely to be the focus. It is difficult to discuss specific areas here as present programmes are still primarily concerned with fundamental understanding. Applications will have to wait on that understanding.

Reference

Bednorz, J. G. & Müller, K. A. 1986 *Z. Phys.* B **64**, 189.

Deltahedral views of fullerene polymorphism

By Donald L. D. Caspar
Rosenstiel Basic Medical Sciences Research Center and Department of Physics, Braideis University, Waltham, Massachusetts 02254-9110, U.S.A.

Fullerenes and icosahedral virus particles share the underlying geometry applied by Buckminster Fuller in his geodesic dome designs. The basic plan involves the construction of polyhedra from 12 pentagons together with some number of hexagons, or the symmetrically equivalent construction of triangular faceted surface lattices (deltahedra) with 12 five-fold vertices and some number of six-fold vertices. All the possible designs for icosahedral viruses built according to this plan were enumerated according to the triangulation number $T = (h^2+hk+k^2)$ of icosadeltahedra formed by folding equilateral triangular nets with lattice vectors of indices h, k connecting neighbouring five-fold vertices. Lower symmetry deltahedra can be constructed in which the vectors connecting five-fold vertices are not all identical. Applying the pentagon isolation rule, the possible designs for fullerenes with more than 20 hexagonal facets can be defined by the set of vectors in the surface lattice net of the corresponding deltahedra. Surface lattice symmetry and geometrical relations among fullerene isomers can be displayed more directly in unfolded deltahedral nets than in projected views of the deltahedra or their hexagonally and pentagonally facted dual polyhedra.

1. Introduction

Buckminster Fuller (1963) called his discipline 'comprehensive anticipatory design science'. Anticipatory science involves recognizing evident answers to questions that have not yet been asked. Fuller's dymaxion geometry (cf. Marks 1960) started with his rediscovery of the cuboctahedron as the coordination polyhedron in cubic close packing, which he renamed the 'vector equilibrium'. Visualizing this figure not as a solid but as a framework of edges connected at the vertices, he transformed the square faces into pairs of triangles to form an icosahedron; and subtriangulation of the spherical icosahedron led to his frequency modulated geodesic domes. Some of Fuller's icosageodesic designs have been used for centuries in the Far East for weaving coolie hats (cf. Pawley 1962). Also anticipating geodesic dome designs, a complete enumeration of all possible subtriangulated icosahedral surface lattices (including chiral plans not used by Fuller) had been described by Goldberg (1937) as a mathematical curiosity. Inspired by Fuller's dome designs, this enumeration was discovered again to explain why isometric virus particles have icosahedral symmetry (Caspar & Klug 1962). Following Fuller's packing notions, Mackay (1962) arranged spherical particles on nested icosahedral surface lattice nets in a non-crystallographic packing which anticipated quasi-crystals. When Kroto *et al.* (1985) discovered C_{60}, which they modelled as the archimedean truncated icosahedron – the familiar football shape and Fuller's lowest frequency modulated icosageodesic sphere – they appropriately named this anticipated structure 'buckminsterfullerene'.

Phil. Trans. R. Soc. Lond. A (1993) **343**, 133–144

Printed in Great Britain

The essential characteristic of fullerenes is that each carbon atom is bonded to three neighbours forming a polyhedral shell with 12 pentagonal and some number of hexagonal facets; for C_{60}, there are 20 hexagonal facets. Possible isomers for higher fullerenes have been enumerated by Fowler, Manolopoulos and colleagues (Fowler, this volume; Fowler *et al.* 1991; Manolopoulos 1991; Manolopoulos & Fowler 1992); and some of the higher fullerenes have been isolated and spectroscopically characterized (Diederich *et al.* 1991; Ettl *et al.* 1991; Diederich & Whetten 1992; Kikuchi *et al.* 1992; Taylor *et al.* 1992). The predicted and observed most stable isomers obey the isolated pentagon rule: no pentagon shares an edge with another pentagon.

The geometry of fullerenes has been conventionally represented by drawings or models of polyhedra constructed from hexagons and pentagons. This geometry can be equally represented by the dual polyhedra: each three-connected vertex corresponds to a triangular facet, and each pentagon or hexagon to a five- or six-connected vertex. Polyhedra constructed from equilateral triangles are called deltahedra (Cundy & Rollet 1954). The particular class of deltahedra consisting of 12 five-vertices (V_5) and some number of six-vertices (V_6) was analysed to account for the symmetry of icosahedral viruses and to predict polymorphic forms of assembly for the coat proteins of these viruses (Caspar & Klug 1962, 1963). Possible designs for deltahedra can be systematically explored by considering the ways in which a plane equilateral triangular net can be cut and folded to form a polyhedron. A utility of the deltahedral representation of non-icosahedral surface lattices, such as many interesting higher fullerenes, is that the symmetry relations of the component trigonal units can be clearly visualized in the unfolded deltahedral lattice net.

2. Quasi-equivalence revisited

Quasi-equivalence was conceived to describe ways in which large numbers of identical protein subunits could build closed containers of predetermined size such as virus capsids by a 'self-assembly' process (Caspar & Klug 1962). Self-assembly presumes specificity of bonding among the structural units. If the same contacts between neighbouring units were used over and over again, in exactly the same way, identical units would be equivalently related and the completed assembly would have some kind of well-defined symmetry: this idea led to the prediction that rod-shaped viruses should have helical symmetry, and 'spherical' viruses might have tetrahedral, octahedral, or icosahedral symmetry (Crick & Watson 1956). Icosahedral symmetry, which was recognized by X-ray crystallography to be the underlying plan for some small isometric virus particles (Klug & Caspar 1960), requires 60, and only 60, equivalent chiral subdivisions. By 1962 chemical studies on two small icosahedral viruses had indicated a count of more than 60 identical protein molecules; and electron microscopy of some larger icosahedral viruses had revealed regular surface arrays of morphological units which were not multiples or submultiples of 60. These observations posed two interrelated questions. Why is icosahedral symmetry preferred? What are the possible designs for an icosahedral shell constructed by regular bonding of a multiple of 60 chiral structural subunits? Anticipation of the answers to these questions was critical to their formulation. The analogy drawn between icosahedral virus particle architecture and Buckminster Fuller's frequency modulated icosageodesic domes (Marks 1960) was the anticipatory key.

Fuller's dome designs involve the subdivision of the surface of the sphere into

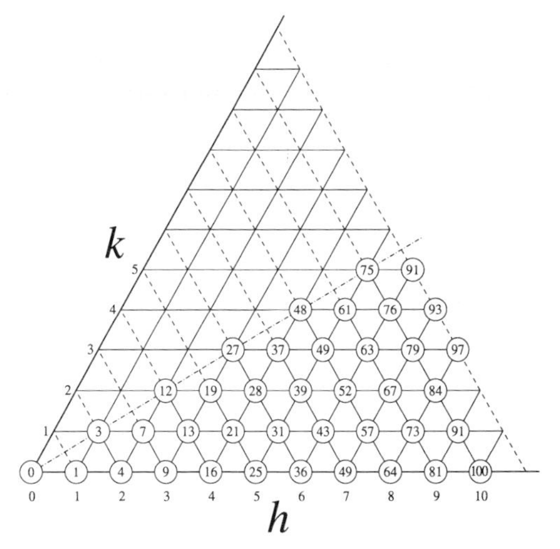

Figure 1. Triangulation numbers $T = (h^2+hk+k^2)$ represented on an equilateral triangular net. An icosadeltahedron (see figure 2) with a five-fold vertex at the origin of this net and a neighbouring five-fold vertex at a lattice point of index h, k will have $\Delta = 20T$ triangular facets, $V_6 = 10(T-1)$ six-connected vertices, and $V_5 = 12$ five-vertices.

nearly equivalent facets arranged with icosahedral symmetry. If identical protein molecules were similarly packed in each geodesic facet, they would then be quasi-equivalently related. Geometrically, the quasi-equivalence could be measured from the variations in the size and shape of the nearly equal geodesic facets; at the molecular level, the same contact points between protein subunits could be used over and over again, but the units themselves or the bonds between them would have to be deformed in slightly different ways in the symmetrically distinct but quasi-equivalent positions. Arrangements that minimize the variation in subunit conformation and bonding should represent minimum energy designs for proteins selected to self-assemble in closed shells containing many molecules.

Possible designs for closed containers built of units that maintain the same pattern of nearest neighbour contacts can be represented by folding plane lattices into polyhedra in various ways, conserving the edge-to-edge connections of the plane facets. Pawley (1962) had shown that only plane lattices with square or equilateral triangular facets, having four- or six-fold rotational symmetry respectively, can be regularly folded onto the surface of convex polyhedra. It is evident that the smallest range of variation in dihedral angles is obtained if a six-connected vertex, V_6, of the equilateral triangular net is transformed into a five-connected vertex, V_5. If only V_5s are allowed on folding the triangular net into a deltahedron, then by Euler's law, $V_5 = 12$ and the number of facets $\Delta = 20+2V_6$. The range of quasi-equivalent variation in the dihedral angles of such a deltahedron will be a minimum if the 12 V_5s are all equivalently related, which requires icosahedral symmetry. This reasoning appeared to account for the selective advantage of icosahedral surface lattices for the construction of virus capsids from some large number of identical protein subunits (Caspar & Klug 1962).

All the possible icosahedral surface lattice designs were enumerated by counting the ways in which the equilateral triangular net could be folded into polyhedra with icosahedral symmetry (called 'icosadeltahedra'). The vector between a neighbouring pair of V_5s of any icosadeltahedron must be a lattice vector of the triangular net.

Since the 12 V_5s are equivalent, the indices (h,k) of the vectors between a lattice point chosen as origin and any point of index (h,k) define all possible icosahedral surface lattices (figure 1). This way of counting is complete and non-redundant (Goldberg 1937; Caspar & Klug 1963). The number of triangular facets in an icosadeltahedron is $\Delta = 20T$, where the triangulation number $T = (h^2+hk+k^2)$, and the number of $V_6 = 10(T-1)$, or $V_5+V_6 = 10T+2$. If $h > k \neq 0$, the icosadeltahedron is chiral, and the pair of vectors (h,k) and (k,h) correspond to enantiomorphs. Figure 2 illustrates models of icosadeltahedra for the first five possible triangulation numbers ($T = 1,3,4,7,9$) built in 1962 from Geodestix components designed by Buckminster Fuller. As models for the design of virus capsids, each deltahedral facet could correspond to a quasi-symmetric trimer of identical, enantiomorphic protein molecules. As models for fullerenes, each deltahedral facet could represent a trivalent carbon atom with quasi or strict three-fold symmetry.

Protein molecules designed to self-assemble into icosahedral capsids of predetermined size may assemble into polymorphic surface lattices of lower symmetry. Some virus capsid proteins may form tubular structures which have a diameter and surface structure similar to the icosahedral particles. Figure 3 illustrates how the $T = 4$ icosadeltahedron of figure 2, with $\Delta = 80$ and point group symmetry I_h, could be transformed into a $\Delta 80$ with D_{5h} symmetry by dividing the structure into two halves perpendicular to a five-fold axis and rejoining after rotating by one unit vector. The deltahedral cap can also be extended by adding rings of V_6 connectors to form a tube of any length. The tube design can be defined by the indices h,k of the circumferential vector; for the tube shown in figure 3, the indices $h,k = 10,0$.

The design of the elongated heads of T-even bacteriophage is based on a chiral deltahedral surface lattice with D_5 symmetry. Mutants of these bacteriophage produce a wide variety of tubular structures built from hexamers of the major capsid protein arranged in cylindrical surface lattices (Yanagida *et al.* 1970). Under some conditions multilayered polyheads are formed, where an innermost tube of diameter *ca.* 40 nm appears to nucleate an assembly of successive layers. The inner tubes have somewhat variable diameters with circumferential vectors mostly within a narrow range of indices $h,k \approx 10,6$–$12,7$. Morphogenesis of the T-even bacteriophage head is a complex process, involving a number of structural and regulatory proteins (Black *et al.* 1992), but the assembled structures have quite regular surface lattice designs.

Considering the chemical complexity of even the simplest icosahedral viruses, it is remarkable that the icosadeltahedral surface lattices representing quasi-equivalent packing of identical molecules accounts so well for the morphology of such a wide variety of structures. The prediction of the quasi-equivalence theory (Caspar & Klug 1962) that the regular icosahedral virus capsids could be built of $60T$ identical protein molecules connected so as to form 12 pentamers and $10(T-1)$ hexamers in an icosahedral surface lattice has been definitely established for the class of viruses with triangulation number $T = 3$, some of whose atomic structures have been solved by X-ray crystallography (Rossmann & Johnson 1989; Harrison 1991).

Larger icosahedral viruses that have been structurally well characterized do not obey the simple quasi-equivalence rule. For example, adenovirus capsids, for which $T = 25$, are built of 240 hexons (six-coordinated units) that are trimers of the major structural protein, and the 12 pentons consist of a different protein (Burnett 1984). Polyomavirus capsids, for which $T = 7$, are built of a single major structural protein,

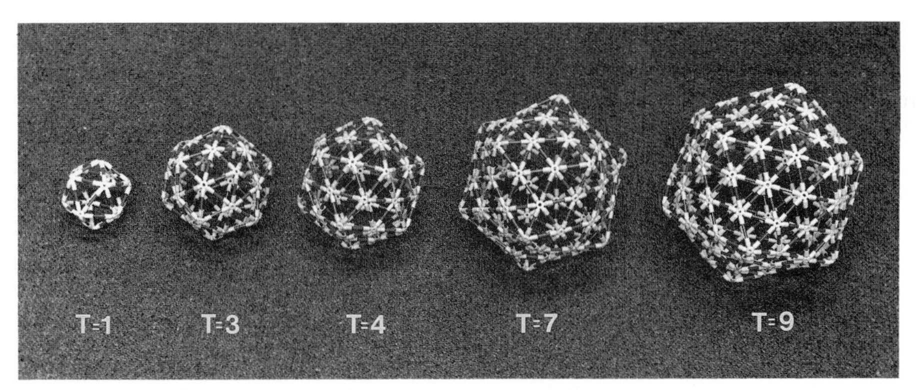

Figure 2. Models of icosadeltahedra for the first five possible triangulation numbers (built from Geodestix components).

Figure 3. Polymorphism of deltahedral surface lattices. The $T = 4$ icosadeltahedron at the left ($\Delta80$, point group I_h) is transformed to a $\Delta80$ with D_{5h} symmetry (middle). At the right, the half-icosahedral cap defined by the $h, k = 10, 0$ circumferential vector has been extended by adding rings of 10 V_6 connectors. The bottom of this tube could be capped symmetrically (as for the $\Delta80$ models) or asymmetrically using the $h, k = 10, 0$ cap shown in figure 4 or the two other $h, k = 10, 0$ caps listed in table 1.

but the 60 hexavalent units are pentamers, chemically identical to the 12 pentavalent pentamers (Rayment *et al.* 1982). Packing of the pentamers in hexagonal slots has been explained by determination of the molecular structure of the simian polyomavirus capsid (Liddington *et al.* 1991), which revealed a highly adaptable protein molecule with extended arms that can form regular contacts in symmetrically very different environments. The adaptability of such versatile protein molecules (Harrison 1991; Caspar 1992) goes beyond the rather modest degree of adjustment postulated for quasi-equivalently connected protein subunits of simple icosahedral virus capsids.

Carbon, although remarkable in the variety of structures that it can form, displays modest adjustability in the configuration of its bonds. Trivalent carbon in graphite can have D_{3h} symmetry, but in fullerene shells only C_{3v}, C_3, C_{1v} and C_1 symmetry are possible and the range of variation in bond distances and angles are limited. Thus, trivalent carbon atoms in fullerene shells conform to the geometrical postulates of the quasi-equivalence theory and the arrangement of the carbon atoms can be represented by the surface lattice of facets of a corresponding deltahedron. Any deltahedron built with 12 V_5s represents a possible fullerene shell and those with all vectors between neighbouring V_5 pairs equal to or greater than $h, k = 1, 1$ satisfy the isolated pentagon rule. Analysis of the possible design of the higher fullerenes by Manolopoulos & Fowler (1992) has revealed a complex polyhedral stereochemistry which can be illustrated by deltahedral surface lattice nets.

Considering the polymorphism possible for units that could form icosahedral surface lattices, it was evident in 1962 (figure 3) that any icosadeltahedron could be divided into two equal halves in three different ways defined by a circumferential vector about an axis of five-, three- or two-fold symmetry, and that additional V_6s could be added to form cylindrical extensions. Furthermore, it was found that any cylindrical lattice formed from the equilateral triangular net (of unit edge length) with a circumferential vector of length $(h^2+hk+k^2)^{\frac{1}{2}}$ equal or greater than that for $h, k = 5, 0$ could be capped with six V_5s. Thus, possible non-icosahedral deltahedra could be identified by cataloging the lattice vectors among six V_5s that could cap each circumferential vector and connecting a pair of caps to a corresponding cylindrical section. In this way, deltahedra with no rotational symmetry were constructed, but 30 years ago the systematic enumeration of the myriad of designs for deltahedra with 12 V_5s and large numbers of V_6s appeared to be an unrewarding exercise. Discovery of the higher fullerenes has now made this classification an enlightening investigation.

3. Deltahedra unfolded

Just as any cylindrical surface lattice formed from the triangular net with a circumferential vector for which $(h^2+hk+h^2) \geqslant 25$ can be capped with six V_5s, so too can any deltahedron with 12 V_5s be unfolded along the direction of a circumferential vector that delimits a pair of six V_5 caps. Figure 4 illustrates the unfolding and refolding of the chiral Δ76 deltahedron in the direction of the $h, k = 10, 0$ circumferential vector. This Δ76 with D_2 symmetry corresponds to the chiral C_{76} isolated and spectroscopically characterized by Ettl *et al.* (1991), and their notation for identifying the 19 carbon atoms in the asymmetric unit is followed in figure 4. The differences in the environments of the facets which distinguish the pyracylene, corrannulene and pyrene sites can be identified more readily in the unfolded net than in pictures of polyhedral models.

Unfolding Δ76 D_2 along the $h, k = 10, 0$ circumferential vector is not unique. In figure 5, the 11 symmetrically distinct ways of unfolding Δ76 D_2 are illustrated, together with the six ways of unfolding the tetrahedrally symmetric Δ76 T_d isomer. In this figure, the boundaries of the surface lattice net are marked by the vectors between nearest neighbour V_5s, rather than along the edges of triangular facets, as in figure 4.

Relations among possible fullerene surface lattices categorized by Manolopoulos & Fowler (1992) can be illustrated by deltahedral nets. For example, any deltahedron can be subtriangulated by applying any triangulation number (figure 1) to increase the number of facets by the factor T. Triangulating by $T = 3$ corresponds to Fowler's leap-frog rule. Any deltahedron circumscribed by a circumferential vector h, k can be

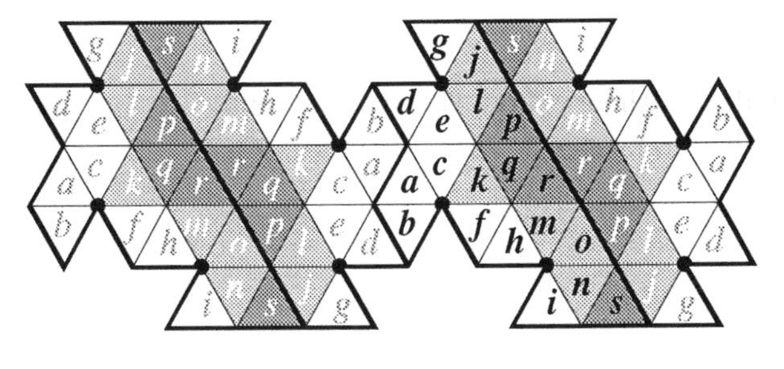

$\Delta 76 \quad D_2 \qquad h,k = 10,0$

Figure 4. Deltahedral surface lattice for the chiral $\Delta 76$ with D_2 symmetry, illustrated for the $h, k = 10, 0$ circumferential vector. An asymmetric unit of this surface lattice, consisting of the 19 facets labelled a–s, has been marked with a bold outline. The labelling of the facets follows the notation of Ettl *et al.* (1991) for the carbon atoms of the chiral C_{76}. Facets a–i (white) correspond to pyracylene sites, j–q (light grey) to corrannulene sites, and p–s (dark grey) to pyrene sites. The three classes of two-fold axes are located between the a–a, r–r, and s–s facets. The opened-out lattice (top) is folded into $\Delta 76$ D_2 by connecting the pair of a facets related by the $h, k = 10, 0$ circumferential vector to form a tube segment which is capped by forming the V_5s marked by black circles. Two views of the half-capped tubular segment are shown with their mirror images (bottom). Folding the surface lattice inside-out would have produced the mirror-image structure.

elongated following the idea of Fowler's cylinder extension rule, in steps of n added V_6s (or $2n\Delta$s) where n is the common factor of h and k. Furthermore, for circumferential vectors with a common factor higher than the axial symmetry of the caps, twisting the caps about the cylinder axis can generate a different deltahedron.

Some of the relations among the 11 smallest deltahedra that obey the isolated pentagon rule (IPΔs) are classified in figure 6. For $\Delta 60$–$\Delta 76$, all circumferential vectors are listed that divide each deltahedron into two caps with six V_5s each. For the five isomers of $\Delta 78$, only those representations of the unfolded deltahedron nets are listed that can be derived by cylindrical extension or alternate combination of the caps of a smaller deltahedron. In general, a tube defined by a circumferential vector h, k can be capped in more than one way. A more complete listing of possible cap designs for tubes larger than those included in table 1 could be produced by enumerating the circumferential vectors for the higher fullerenes catalogued by Manolopoulos & Fowler (1992).

Of the deltahedra listed in figure 6, $\Delta 60$ I_h, $\Delta 72$ D_{6d}, and $\Delta 78$ D_{3h} can be respectively derived by the $T = 3$ triangulation as follows: $\Delta 60$ from $\Delta 20$ the

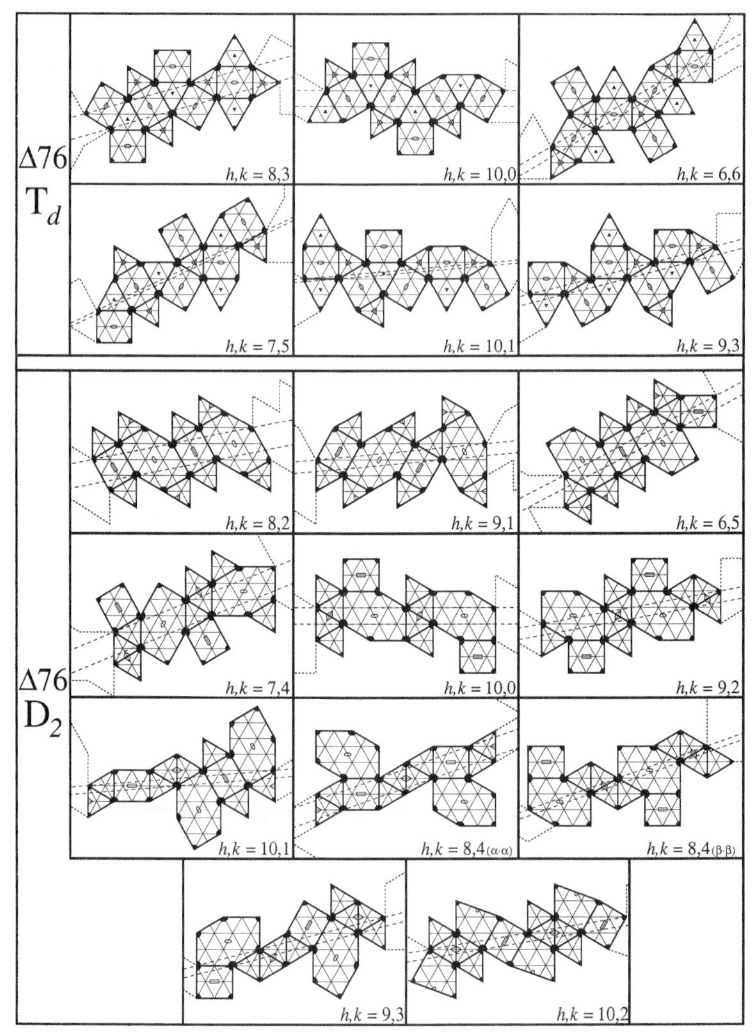

Figure 5. Unfolding of the tetrahedral T_d and chiral D_2 isomers of Δ76 according to the possible circumferential vectors that divide the deltahedra into two caps. The pair of circumferential vectors at the boundary of each tubular segment (which contains no V_5s) are marked by dashed lines. V_5 lattice points are indicated by black circular sectors and the vectors between nearest neighbour V_5s are marked by bold lines. The dotted lines indicate how the two ends of the circumferential vectors are joined to form the tubular segment. Folding only along the vectors between V_5s would produce a polyhedron with the marked polygons as facets; whereas folding only along edges of the triangular net produces the deltahedron (cf. figure 4). The symmetrically distinct rotational axes for the T_d and D_2 point groups are marked with distinguishing shapes. Circumferential vectors have been chosen with $h > k$ and are arranged in order of increasing length $(h^2+hk+k^2)^{\frac{1}{2}}$. (Interchanging h and k produces the enantiomorphic net for the D_2 surface lattices.) Increasing the separation of the pairs of circumferential vectors by addition of V_6s leads to cylinder extension generating higher deltahedra (cf. table 1).

icosahedron; Δ72 from Δ24, the hexagonal antiprism (which can be derived from the icosahedron by adding two V_6s in extension along a two-fold axis); and Δ78 from Δ26, derived from the icosahedron by addition of three V_6s extended along a three-fold axis. In turn, by extension, Δ70 derives from Δ60; Δ74 from Δ70; Δ76 D_2 from Δ60 and Δ72 by one path or from Δ70 by two different paths; and Δ76 T_d from Δ74.

Δn	Point Group	Index *h,k* of Circumferential Vectors																	
		5,5	6,4	7,3	9,0	8,2	9,1	6,5	7,4	8,3	10,0	9,2	6,6	7,5	10,1	8,4	9,3	11,0	10,2
Δ60	I_h	α·α			α·α	α·α													
Δ70	D_{5h}	α·α					α·α		α·α		α·β								
Δ72	D_{6d}					α·α			β·β				α·α		α·α				
Δ74	D_{3h}								α·α	α·α	α·γ			α·α	β·β		α·α		
Δ76	T_d									α·α	α·α		β·β	α·β	β·γ		α·β		
"	D_2					α·α	α·α	α·α	α·α		α·α	α·α			δ·δ	α·α β·β	γ·γ		α·α
Δ78	D_{3h}				α·α				β·β						α·α				
"	C_{2v}						α·α		α·β										
"	$C_{2v'}$								α·α										
"	$D_{3h'}$									α·α				β·β	γ·γ		β·β		
"	D_3				α·α														

Figure 6. Circumferential vectors of deltahedra with isolated pentameric vertices (IPΔs). All circumferential vectors from $h, k = 5, 5$ to $10, 2$ are listed in order of increasing length $(h^2+hk+k^2)^{\frac{1}{2}}$. This listing indicates possible designs for fullerene tubes that can be capped with isolated pentamers. No such caps are possible for circumferential vectors $h, k = 6, 4$ or $7, 3$; for longer circumferential vectors, more than one arrangement of six V_5s may cap a tube segment. Symmetrically distinct caps are sequentially designated α, β, γ, δ, ..., as they are listed for each h, k vector for progressively larger IPΔs. Surface lattice nets for the Δ76 and Δ78 isomers are illustrated in figures 5 and 6 respectively.

Furthermore, as is evident from figures 4 and 5,, the chiral Δ76 D_2 can be converted into the tetrahedral Δ76 T_d by shifting the top cap related by the $10, 0$ circumferential vector one lattice unit to the left. Shifting two units to the left generates the D_2 enantiomer. Continuing this shifting three or four steps brings a V_5 pair one lattice unit apart, which violates the isolated pentagon rule, and five steps comes back by symmetry to a starting point. Cylinder extension, by adding V_6s, can also lead to apposition of V_5s that violates the isolated pentamer rule, indicated for example by the blank entries for Δ72 and Δ74 in the column for the $9, 1$ circumferential vector in figure 6.

The interrelation of the five isomers of Δ78 is illustrated in figure 7. At the left, unfolded surface lattices are drawn as in figure 5 for selected circumferential vectors from the listing in figure 6. In the centre, projected views of the fullerene polyhedra are compared with their deltahedra duals in the same orientation. At the right, surface lattices for the four mirror-symmetric Δ78 isomers are drawn with boundaries marked along the edges of triangular facets as in figure 4. These unfolded nets emphasize the invariant and variable aspects of the polymorphic interchange among the isomers of C_{78} $D_{3h} \rightleftharpoons C_{2v} \rightleftharpoons C_{2v'} \rightleftharpoons D_{3h'}$ as described by Diederich *et al.* (1991). Switching among these isomers by rotation of a C_2-unit in a pyracylene rearrangement, following the mechanism proposed by Stone & Wales (1986), is illustrated in figure 7 by the step-wise reorientation of the three shaded pairs of triangular facets. This local reorientation leads to the interchange of a V_5 and V_6 pair in the surface lattice. In contrast, there is no simple interchange between the chiral D_3 isomer and any of the four mirror-symmetric isomers. The Δ78 D_{3h} can be transformed to the D_3 isomer shown at the top of figure 7 by shifting the upper cap defined by the $9, 0$ circumferential vector one unit to the right, or to its enantiomer by the opposite shift. However, for C_{78} this would correspond to the improbable breaking and reforming of 18 C–C bonds.

NMR spectra of purified C_{78} obtained in different laboratories (Diederich *et al.* 1991;

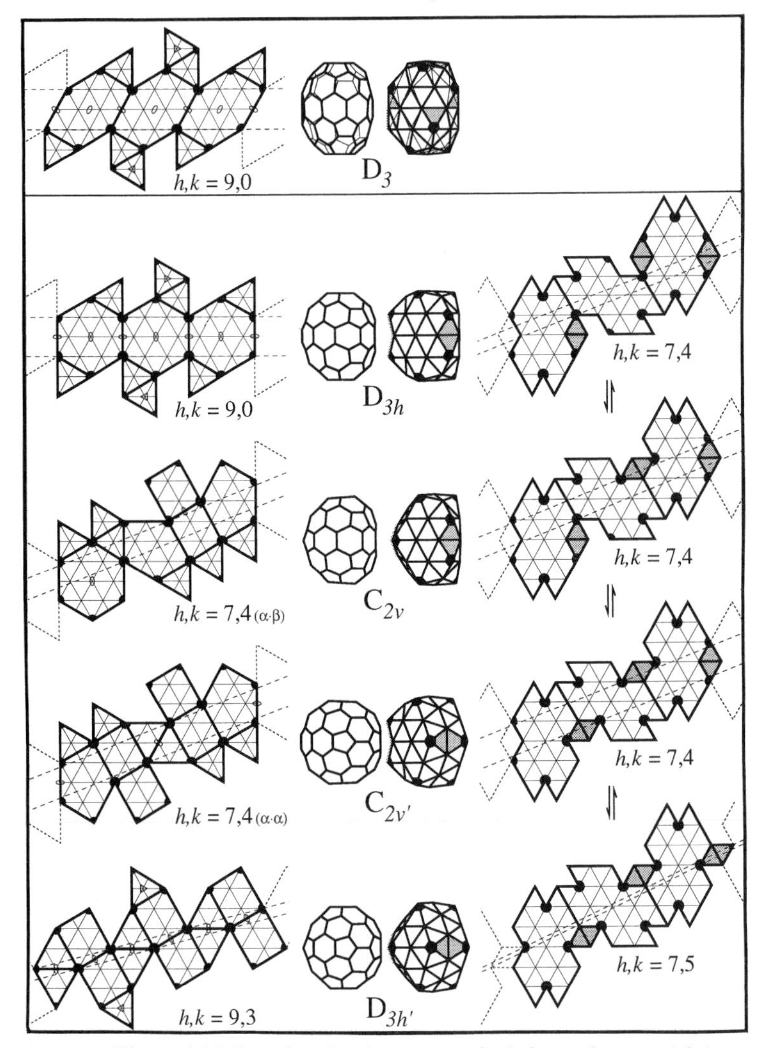

Figure 7. Δ78 isomers. The unfolded surface lattice nets at the left are drawn with boundaries along the vectors between nearest neighbour V_5s which are marked by the black circular sectors, whereas the boundaries for the nets at the right are along the edges of deltahedral facets. The projected views of the fullerene polyhedra and deltahedra duals in the centre column are all oriented with a corresponding two-fold axis horizontal. For the four mirror-symmetric isomers, there is one mirror plane in the plane of projection and an orthogonal horizontal one. Marking the symmetry elements for each isomer on the deltahedral surface lattice net defines the asymmetric unit.

Kikuchi *et al.* 1992; Taylor *et al.* 1992) indicate variable proportions of the isomers D_3, C_{2v} and $C_{2v'}$. The spectra expected for each isomer can be predicted by enumerating the number of symmetrically distinct carbon atoms in the asymmetric unit and noting their local environment (Fowler *et al.* 1991). The deltahedral facets composing the asymmetric units of each Δ78 isomer in figure 7 could be labelled as in figure 4, distinguishing the pyracylene, corrannulene and pyrene sites.

The unfolded deltahedral nets provide a convenient way to illustrate the environments of the different carbon atoms in the asymmetric unit of higher fullerenes, in particular those of relatively low symmetry. The interrelation among fullerenes of different size and among various isomers can also be displayed by such nets. In particular, interconversion pathways that involve local reorientations at relatively distant sites can be simply mapped. Furthermore, the classification of the deltahedra according to their circumferential vectors provides a systematic way to enumerate how nanotubes can be capped, and how multilayered tubes (Iijma 1991) or shells (McKay *et al.* 1992) can be successively encapsulated.

I thank Eric Fontano for preparing the computer graphics diagrams and Marie Craig for photographic assistance. This work has been supported by United States Public Health Service Grant CA47439 from the National Cancer Institute.

References

Black, L. F., Showe, M. K. & Steven, A. C. 1992 Morphogenesis of the T4 head. In *Bacteriophage T4* (ed. T. Karam & F. A. Eiserling), 2nd edn. Washington, D.C.: American Society for Microbiology.

Burnett, R. M. 1984 Structural investigations of hexon, the major coat protein of adenovirus. In *Biological macromolecules and assemblies* (*Volume 1: Virus structures*) (ed. F. A. Jurnak & A. McPherson), pp. 377–385. New York: John Wiley & Sons.

Caspar, D. L. D. 1992 Virus structure puzzle solved. *Current Biol.* **2**, 169–171.

Caspar, D. L. D. & Klug, A. 1962 Physical principles in the construction of regular viruses. *Cold Spring Harbor Symp. Quant. Biol.* **27**, 1–24.

Caspar, D. L. D. & Klug, A. 1963 Structure and assembly of regular virus particles. In *Viruses, nucleic acids, and cancer*, pp. 27–39. Baltimore: Williams & Wilkins.

Crick, F. H. C. & Watson, J. D. 1956 The structure of small viruses. *Nature, Lond.* **177**, 473–475.

Cundy, H. M. & Rollett, A. P. 1954 *Mathematical Models*, 2nd edn. Oxford: Clarendon Press.

Diederich, F., Whetten, R. L., Thilgen, C., Ettl, R., Chao, I. & Alvarez, M. M. 1991 Fullerene isomerism: Isolation of C_{2v}–C_{78} and D_3–C_{78}. *Science, Wash.* **254**, 1768–1770.

Ettl, R., Chao, I., Diederich, F. & Whetten, R. L. 1991 Isolation of C_{76}, a chiral (D_2) allotrope of carbon. *Nature, Lond.* **353**, 149–153.

Fowler, P. W., Batten, R. C. & Manolopoulos, D. E. 1991 The higher fullerenes: a candidate for the structure of C_{78}. *J. chem. Soc. Faraday Trans.* **87**, 3103–3104.

Fuller, R. B. 1963 *Ideas and integrities.* Englewood Cliffs, New Jersey: Prentice-Hall.

Goldberg, M. 1937 A class of multi-symmetric polyhedra. *Tôhoku Math. J.* **43**, 104–108.

Harrison, S. C. 1991 What do viruses look like? *Harvey Lectures* **85**, 123–148.

Iijma, S. 1991 *Nature, Lond.* **354**, 56–58.

Kikuchi, K., Nakahara, N., Wakabayashi, T., Suzuki, S., Shiromaru, H., Miyake, Y., Saito, K., Ikemoto, I., Kainosho, M. & Achiba, Y. 1992 NMR characterization of isomers of C_{78}, C_{82} and C_{84} fullerenes. *Nature, Lond.* **357**, 142–145.

Klug, A. & Caspar, D. L. D. 1960 The structure of small viruses. *Adv. Virus Res.* **7**, 225–325.

Kroto, H. W., Heath, J. R., O'Brien, S. C., Curl, R. F. & Smalley, R. E. 1985 C_{60}: Buckminsterfullerene. *Nature, Lond.* **318**, 162–164.

Liddington, R. C., Yan, Y., Moulai, J., Sahli, R., Benjamin, T. L. & Harrison, S. C. 1991 Structure of simian virus 40 at 3.8-Å resolution. *Nature, Lond.* **354**, 278–284.

Mackay, A. L. 1962 A dense non-crystallographic packing of equal spheres. *Acta crystallogr.* **15**, 916–918.

Manolopoulos, D. E. 1991 Proposal of a chiral structure for the fullerene C_{76}. *J. chem. Soc. Faraday Trans.* **87**, 2861–2862.

Manolopoulos, D. E. & Fowler, P. W. 1992 Molecular graphs, point groups, and fullerenes. *J. chem. Phys.* **96**, 7603–7614.

Marks, R. W. 1960 *The dymaxion world of buckminster fuller*. New York: Reinhold.

McKay, K. G., Kroto, H. W. & Wales, D. J. 1992 *J. chem. Soc. Faraday Trans.* **88**, 2815–2821.

Pawley, G. S. 1962 Plane groups on polyhedra. *Acta crystallogr.* **15**, 49–53.

Rayment, I., Baker, T. S., Caspar, D. L. D. & Murakami, W. T. 1982 Polyomavirus capsid structure at 22.5 Å resolution. *Nature, Lond.* **295**, 110–115.

Rossmann, M. G. & Johnson, J. E. 1989 Icosahedral RNA virus structure. *A. Rev. Biochem.* **58**, 533–573.

Stone, A. J. & Wales, D. J. 1986 Theoretical studies of icosahedral C_{60} and some related species. *Chem. Phys. Lett.* **128**, 501–503.

Taylor, R., Langley, G. J., Dennis, T. J. S., Kroto, H. W. & Walton, D. R. M. 1992 A mass spectrometric-NMR study of fullerene-78 isomers. *J. chem. Soc. Chem. Commun.*, 1043–1046.

Yanagida, M., Boy de la Tour, E., Alff-Steinberger, C. & Kellenberger, E. 1970 Studies on the morphopoiesis of the head of bacteriophage T-even. VII. Multilayer polyheads. *J. molec. Biol.* **50**, 35–58.

Geodesic domes and fullerenes

By T. Tarnai

Department of Mechanics, Faculty of Civil Engineering, Technical University of Budapest, Budapest, Müegyetem rkp 3, H-1521 Hungary

The structural form of geodesic domes, composed of pentagons and hexagons, played an important role in understanding the structure of carbon clusters. In this paper an analogy between geodesic domes and fullerenes is investigated. A brief survey is given of the geometry of geodesic domes applied in engineering practice, in particular of the geodesic domes bounded by pentagons and hexagons. A connection is also made between these sorts of geodesic domes and the mathematical problem of the determination of the smallest diameter of n equal circles by which the surface of a sphere can be covered without gaps. It is shown that the conjectured solutions to the sphere-covering problem provide topologically the same configurations as fullerene polyhedra for some values of n. Mechanical models of fullerenes, composed of equal rigid nodes and equal elastic bars are also investigated, and the equilibrium shapes of the space frames that model C_{28}, C_{60} and C_{240} are presented.

1. Introduction

From visual inspection one can easily discover an analogy between the structure of C_{60} and the inner layer of the structure of the great U.S. pavilion of R. B. Fuller at the 1967 Montreal Expo. This analogy and other geodesic structures of Fuller were responsible for the name of C_{60}: Buckminsterfullerene (Kroto *et al.* 1985). This is not the first time that Fuller's geodesic domes have helped researchers to understand the structure of matter. In the early 1960s Fuller's geodesic domes, especially his tensegrity spheres, inspired Caspar & Klug (1962) to develop the principle of quasi-equivalence in virus research.

The aim of this paper is to investigate the analogy between geodesic domes bounded by pentagons and hexagons and fullerenes. It is possible to construct a network of this sort of geodesic dome by covering the sphere with circles (Tarnai & Wenninger 1990; Pavlov 1990). We shall consider the mathematical problem of determining of the smallest diameter of n equal circles by which the surface of a sphere can be covered without gaps, and shall show that the conjectured solutions to the sphere-covering problem provide topologically the same configurations as fullerene polyhedra for some n.

An easy way of visualizing the structure of fullerenes is to make a physical model, for example, to assemble equal angle planar trivalent connectors and equal length plastic tubes. Mechanically, the polyhedron-like structure so obtained can be considered as a space frame with equal rigid nodes and equal elastic bars such that three bars meet and form angles of 120° at each node. The closed net shape arises by deformation of the bars in a state of self-stress. The edges of the polyhedron obtained

Phil. Trans. R. Soc. Lond. A (1993) **343**, 145–154

Printed in Great Britain

Figure 1. Geodesic dome in the town of Baku, Azerbaidzhian, erected in 1976. (With the kind permission of Dr G. N. Pavlov.)

will be curved, not straight. We have made calculations to determine the equilibrium shape of the space frames modelling C_{28}, C_{60} and C_{240}. The main results are presented below.

2. Geodesic domes and spherical honeycombs

Although, to my knowledge, the first geodesic dome was designed by W. Bauersfeld and was constructed in the early 1920s in Germany, geodesic domes are associated with the name Richard Buckminster Fuller. Fuller called a dome geodesic if the lines on the surface of the sphere producing a three-way grid are geodesics, that is, great circles of the sphere. These can be obtained in the easiest way if equal regular triangulation is made on all faces of an icosahedron, and the resulting network is projected from the centre of the icosahedron onto the surface of its circumsphere. Later the term *geodesic dome* was used for all polyhedral structures resulting from triangulation of the spherical surface having icosahedral symmetry where the edge lengths of the triangles in the network do not differ too much from each other. The U.S. pavilion at the 1967 Montreal Expo is considered as one of the best examples of a geodesic dome, however, it is a geodesic dome only in the latter, more general sense, since the lines producing the triangular subdivision on the bottom part of the spherical surface are not great circles but small circles of the sphere. Over the years numerous triangular subdivision methods have been developed (Tarnai 1987, 1990).

If the subdivision frequency along an edge of the icosahedron is divisible by 3 then it is easy to make a hexagonal geodesic network from a triangular one by removing the edges. In contrast to geodesic domes of a triangular network, however, geodesic domes with hexagonal networks as single-layer bar-and-joint structures are not rigid. In engineering, hexagonal networks can be applied for single-layer space frames with rigid nodes (an example of this as a dome made of 'dog-bone' elements can be seen in Fuller (1969)), or for plate structures (figure 1); or for one of the layers of double-layer grids. Goldberg (1937) introduced a classification for hexagonal tessellations on the surface of an icosahedron, which also incorporated skewed arrangements. These have been used intensively in virus research (Caspar & Klug 1962), and this idea has also appeared in the fullerene field (Klein *et al.* 1986). It is also possible to apply Goldberg's skewed arrangements to geodesic domes (Tarnai 1984,

Figure 2. Model of a spherical 'hexagonal' net with equal edges and with icosahedral symmetry.

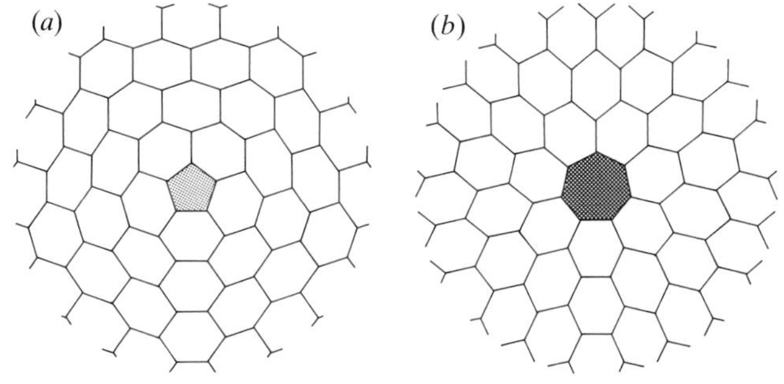

Figure 3. Arrangement of hexagons of equal edge lengths in the plane around an individual (*a*) pentagon with D_5 symmetry and (*b*) heptagon with D_7 symmetry.

1989, Tarnai & Wenninger 1990), but these sorts of domes have not yet been used in practice.

In a triangular subdivision of a spherical surface with icosahedral symmetry – except for the icosahedron itself – the edge lengths in the network are not all equal. A network composed of several hexagons and 12 pentagons can be constructed on the spherical surface with icosahedral symmetry so that the edge lengths are equal (figure 2). In such a network, however, angles of hexagons can differ significantly from the 120° of regular hexagons. In an extreme case, where the common edge length is very small compared with the radius of the sphere, the local spherical networks are approximately planar, and we find that hexagons are elongated in a circumferential direction around the pentagon (figure 3*a*) and that the angles of the hexagons vary between 108° and 144°. If an individual heptagon were to be introduced into a planar 'hexagonal' network it would have the opposite effect: the hexagons would be elongated in the radial direction around the heptagon (figure 3*b*) and the angles of the hexagons would vary between 128.6° and 102.8°. If pentagons and heptagons are introduced together, a network can be obtained where the angles

Figure 4. Polyhedron model of locally minimum covering of the sphere by 16 equal circles.

of hexagons remain close to 120°. In nature there are several examples of spherical honeycombs in which some heptagons are involved (Tarnai 1989). Although I do not know of examples of geodesic domes with 5-, 6-, and 7-gonal faces, numerous examples of geodesic domes, especially radomes, exist with triangular network having 5-, 6-, and 7-valent vertices. A remarkable structure can be seen in Schönbach (1971).

The large hypothetical fullerenes are not spherical (Kroto 1988), so it is probably not necessary to have heptagons in the network to keep bond angles close to 120° and thus minimize the potential energy. It is worth mentioning, however, that Mackay & Terrones (1991) suggested a carbon structure fitted to an infinite periodic minimum surface. Because of the negative gaussian curvature of such a surface, it should contain polygons with more than six sides as well as hexagons.

3. Covering the sphere by circles

The different subdivision procedures worked out for the construction of a geodesic dome usually result in distorted hexagons. Pavlov (1990), however, has developed an efficient method for subdividing a spherical surface, which results in inscribed polyhedra (of a sphere) bounded by plane pentagons and hexagons. The geodesic dome in figure 1 was constructed in this way. The circumscribed circles of the polyhedron faces yield an economical way of covering a sphere. Pavlov's method opened up an interesting possibility for building a geodesic dome: to construct a dome with overlapping circles as structural elements (Tarnai & Wenninger 1990).

The problem of how to economically cover a sphere by circles is well known in geometry. In the simplest case, where all the circles are identical, the problem is as follows (Fejes Tóth 1972): how may a sphere be covered by n equal circles (spherical caps) so that the angular radii of the circles are as small as possible? Mathematically proven solutions are known for $n = 2$–7 and for $n = 10$, 12 and 14; conjectured solutions exist for the other values of n up to 20 and for some sporadic values of $n > 20$.

It turns out that there is a correspondence between the topology of arrangements found in the proven and conjectured solutions of the covering problem and various distinct physical problems. Tarnai & Gáspár (1991) analysed the covering problem

Figure 5. Polyhedron model of locally minimum covering of the sphere by 32 equal circles. (Reprinted from Tarnai & Wenninger (1990) with permission of *Structural Topology* (Montreal) and Dr M. J. Wenniner.)

for small numbers of circles on the basis of the structural form of coated vesicles. Tarnai (1991) pointed out that the coated vesicles identified by Crowther *et al.* (1976) and an additional one having a truncated icosahedron form provide the best covering configurations for $n = 16$, 20 and 32. Interestingly, among the conjectured solutions resulting in local optima, for $n = 16$, 18 and 32 we obtained topologically the same configurations as those of Kroto (1988) published for C_{28}, C_{32}, and C_{60} (Tarnai & Gáspár 1991). The configurations of the conjectured solutions of the covering problem for $n = 16$ and 32 are shown in figures 4 and 5.

Inspection of the structure in figure 5 reveals that the length of an edge separating a pentagon from a hexagon is greater than the length of an edge separating two hexagons. This property of covering the sphere by 32 equal circles (considering only its tendency and not the actual lengths) is consistent with the measurements of C_{60} by Hedberg *et al.* (1991) who found the bond lengths in the five-member rings to be larger than the lengths of bonds fusing the six-membered rings.

4. Stick models of fullerenes

A simple way to appreciate the shape of fullerene is to construct a physical model in which rigid planar trivalent nodal connectors represent the atoms and flexible plastic bars (tubes) of circular cross-section represent the bonds. From a mechanical point of view the model may be considered as a polyhedron-like space frame whose equilibrium shape is due to self-stress caused by deformation of bars. We suppose that the bars are equal and straight in the rest position and that they are inclined relative to each other at every node with angle of 120°. The material of the bars is assumed to be perfectly elastic and that Hooke's law is valid. All the external loads and influences are neglected and only self-stress is taken into account. Then we pose the question: What is the shape of the model subject to these conditions? To answer this question we apply the idea used for coated vesicles by Tarnai & Gáspár (1989).

Let us first investigate this mechanical problem for the model of C_{28} having tetrahedral symmetry (Kroto 1988), and arrange the structural elements in a plane and join the corresponding bar ends and nodes. As a result, a ball-like space frame

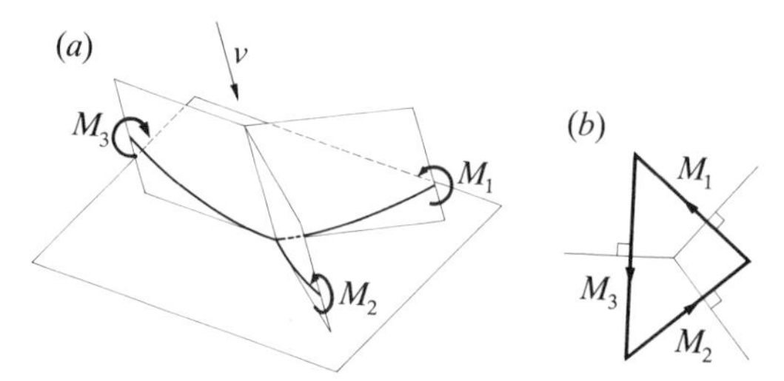

Figure 6. Equilibrium of a node. (*a*) Three bars bending at a node. (*b*) Closed triangle of moment vectors in a plane perpendicular to line v.

is obtained whose bars are no longer straight but curved. Bending the bars generates a state of self-stress in the frame. Due to symmetry, the bars of the frame take on plane-curve form and thus do not twist. This allows us to surmise that the bars of the space frame are under pure bending. The axis of a bar in pure bending takes on the form of a circular arc in the plane of the bend. The bending moments (couples) can be represented by vectors which are perpendicular to the plane of bending.

Let us consider the equilibrium of a node. At a node three bars meet and are subject to bending moments M_1, M_2 and M_3 (figure 6*a*). Under the influence of these moments the bars bend and their axes become circular arcs of radii R_1, R_2 and R_3 respectively, and these circular arcs lie in three different planes. The node is in equilibrium if the vectors of the three bending moments form a closed triangle (figure 6*b*). Since every bending moment vector is perpendicular to its plane, the vector triangle is closed only if the three planes (i.e. the planes of circular arcs) intersect in a common line v. In general, this line v is not perpendicular to the common tangent plane of the curved axes of the three bars at the node, but because of the above-mentioned property, it is perpendicular to the plane of the vector triangle.

The whole space frame is in equilibrium if every node is in equilibrium, i.e. if the bending moment vectors of the bars form a closed polyhedron bounded by triangles. Due to the perpendicularity conditions, the polyhedron formed by the bending moment vectors will be the dual of the curved edge polyhedron. The curved-edged primal polyhedron, i.e. the model itself – due to symmetry – has three different kinds of edges: AB, BC, CC (circular arcs of radii R_1, R_2, R_3) as seen in figure 7*a*, and the dual polyhedron has three different edge lengths: M_1, M_2, M_3 (figure 7*b*). The nodes of the model do not lie on a single spherical surface. In the unit edge length case the radii and the chord lengths of the edges are:

$$R_1 = 1.43760214, \qquad |AB| = 0.979960623,$$
$$R_2 = 1.62321490, \qquad |BC| = 0.984261033,$$
$$R_3 = 2.07219822, \qquad |CC| = 0.990324758.$$

Coordinates of nodes are given in Tarnai & Gáspár (1989).

The C_{60} model has two different kinds of bar forms: AA′ and AA″ (circular arcs of radii R_1 and R_2) as seen in figure 8. In the unit bar length case, the radii and the chord lengths of the curved bar axes are:

$$R_1 = 2.45953818, \qquad |AA'| = 0.993126401,$$
$$R_2 = 2.31046566, \qquad |AA''| = 0.992212954.$$

The model looks like a geodesic dome since all the nodes lie on a single sphere whose radius is, interestingly, equal to R_1. Therefore the curved bars which separate two

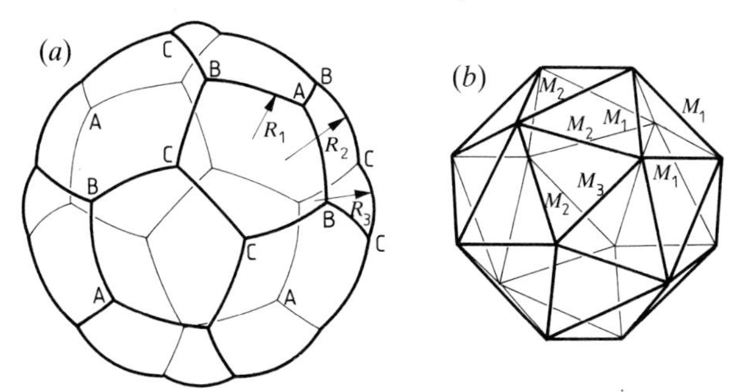

Figure 7. The equilibrium of the space frame modelling C_{28}. (*a*) The primal curved-edged polyhedron formed from bars. (*b*) The dual polyhedron formed from moments.

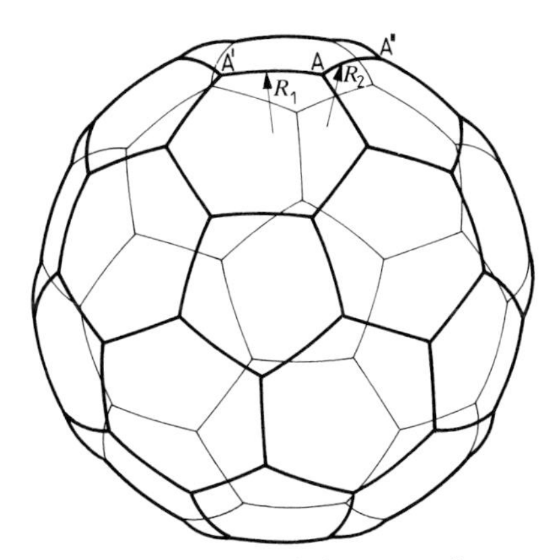

Figure 8. The equilibrium shape of the space frame modelling C_{60}.

adjacent hexagons lie on the same spherical surface. The planes of the curved bars which separate a hexagon and a pentagon do not pass through the centre of the sphere. In this model the chord lengths contrast with the bond lengths measured by Hedberg *et al.* (1991). In the model $|AA'| > |AA''|$, but in C_{60} itself the direction of the inequality is just the opposite: $1.398 < 1.455$.

Kroto & McKay (1988) found that the shapes of stick models of larger icosahedral fullerenes diverge from the geodesic dome form and develop shapes into icosahedral symmetry gradually shifting to the icosahedron as to a polyhedron. Our calculations seem to confirm this trend. The C_{240} model has five different kinds of bar forms. The five bars are shown in a simplified way by thick lines in figure 9 where the large regular triangle composed of dashed lines is a face of the icosahedron. The bars develop pure bending deformations except for bar AC and its symmetric replicas which are also twisted. The unit bar lengths the radii and the chords of the curved bar axes in figure 9 are:

$$
\begin{aligned}
R_1 &= 2.437\,315\,83, & |AB| &= 0.993\,000\,754,\\
R_2 &= 2.294\,060\,14, & |BB'| &= 0.992\,101\,446,\\
R_3 &= 94.375\;\;8743, & |CC'| &= 0.999\,995\,322,\\
R_4 &= 4.730\,430\,37, & |CC''| &= 0.998\,146\,064,\\
& & |AC| &= 0.999\,523\,763.
\end{aligned}
$$

The largest deformations occur in the neighbourhood of the vertices of the

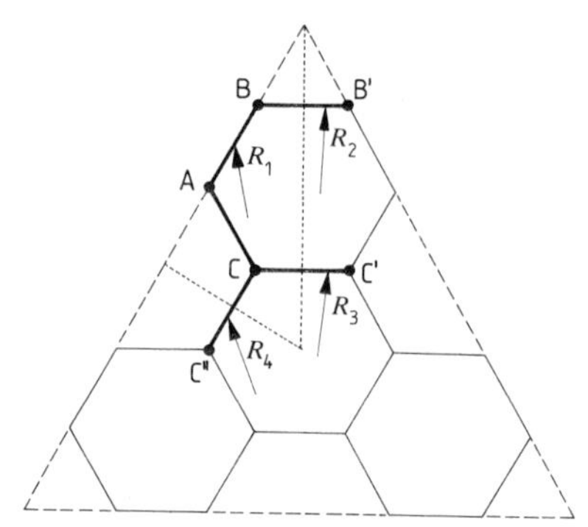

Figure 9. Arrangement of bars of the space frame modelling C_{240} on a face of the icosahedron.

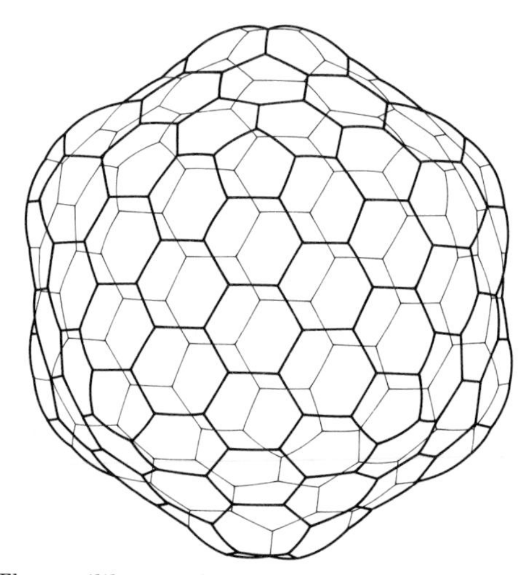

Figure 10. The equilibrium shape of the space frame modelling C_{240}.

icosahedron, and the hexagon in the middle of the faces of the icosahedron becomes flat. For the sake of brevity the coordinates of nodes are not given here; they will be published elsewhere but the complete model is shown in figure 10. Calculations for the model of C_{540} are in progress.

5. Conclusions

The consideration of analogies between geodesic domes and fullerenes is fruitful in both directions. At the very beginning of fullerene research, geodesic dome concepts helped chemists to recognize the structure of these hollow carbon clusters.

In the reverse direction, fullerene studies can indirectly help in the geodesic dome design. The intensive research on fullerenes is providing many suggestions for new structural forms composed of hexagons and pentagons, which may be considered as basic configurations for analysing the fundamental problem of minimum covering of a sphere by circles and also as new configurations for geodesic domes. In this way new geometrical results are also to be expected in the future as a consequence of the detailed studies which are now being carried out on the fullerenes.

I thank Professor Zs. Gáspár for helpful discussions and for making calculations. This work was supported by OTKA I/3 Grant No. 41 awarded by the Hungarian Scientific Research Foundation.

References

Caspar, D. L. D. & Klug, A. 1962 Physical principles in the construction of regular viruses. *Cold Spring Harbor Symp. Quant. Biol.* **27**, 1–24.

Crowther, R. A., Finch, J. T. & Pearse, B. M. F. 1976 On the structure of coated vesicles. *J. molec. Biol.* **103**, 785–798.

Fejes Tóth, L. 1972 *Lagerungen in der Ebene auf der Kugel und im Raum*, 2nd edn. Berlin: Springer-Verlag.

Fuller, R. B. 1969 *Ideas and integrities.* New York: Collier Books.

Goldberg, M. 1937 A class of multi-symmetric polyhedra. *Tôhoku Math. J.* **43**, 104–108.

Hedberg, K., Hedberg, L., Bethune, D. S., Brown, C. A., Dorn, H. C., Johnson, R. D. & de Vries, M. 1991 Bond lengths in free molecules of buckminsterfullerene, C_{60}, from gas-phase electron diffraction. *Science, Wash.* **254**, 410–412.

Klein, D. J., Seitz, W. A. & Schmalz, T. G. 1986 Icosahedral symmetry carbon cage molecules. *Nature, Lond.* **323**, 703–706.

Kroto, H. W. 1988 Space, stars, C_{60} and soot. *Science, Wash.* **242**, 1139–1145.

Kroto, H. W., Heath, J. R., O'Brien, S. C., Curl, R. F. & Smalley, R. E. 1985 C_{60}: Buckminsterfullerene. *Nature, Lond.* **318**, 162–163.

Kroto, H. W. & McKay, K. G. 1988 The formation of quasi-icosahedral spiral shell carbon particles. *Nature, Lond.* **331**, 328–331.

Mackay, A. L. & Terrones, H. 1991 Diamond from graphite. *Nature, Lond.* **352**, 762.

Pavlov, G. N. 1990 Determination of parameters of crystal latticed surfaces composed of hexagonal plane facets. *Int. J. Space Structures* **5**, 169–185.

Schönbach, W. 1971 Als Netzkuppel ausgebildetes Radom mit 49 m Durchmesser. *Der Stahlbau* **40**, 45–55.

Tarnai, T. 1984 Optimization of spherical networks for geodesic domes. In *Proc. Third Int. Conf. on Space Structures* (ed. H. Nooshin), pp. 100–104. Guildford, U.K.: University of Surrey.

Tarnai, T. (ed.) 1987 *Spherical grid structures.* Budapest: Hungarian Institute for Building Science.

Tarnai, T. 1989 Buckling patterns of shells and spherical honeycomb structures. *Computers Math. Applic.* **17**, 639–652.

Tarnai, T. (ed.) 1990 Special issue on geodesic forms. *Int. J. Space Structures* **5**, 155–374.

Tarnai, T. 1991 The observed form of coated vesicles and a mathematical covering problem. *J. molec. Biol.* **218**, 485–488.

Tarnai, T. & Gáspár, Zs. 1989 The shape of coated vesicles. *RAD Jugosl. Akad. Znan, Umj., Math.* (*444*) **8**, 121–130.

Tarnai, T. & Gáspár, Zs. 1991 Covering a sphere by equal circles, and the rigidity of its graph. *Math. Proc. Camb. phil. Soc.* **110**, 71–89.

Tarnai, T. & Wenninger, M. J. 1990 Spherical circle-coverings and geodesic domes. *Structural Topology* **16**, 5–21.

Discussion

S. Iijima (*NEC Corporation, Japan*). I used molecular mechanics to optimize the geometry of giant fullerenes up to C_{960}. Contrary to Professor Kroto's observation, as we optimize the structure approaches a sphere, so the strain is apparently distributed throughout the molecule.

T. Tarnai. A very strong bond deformation should exist in the hexagons of the result is to be spherical.

J. P. Hare (*Sussex University, U.K.*). I have made an icosohedral large fullerene. To make a spherical version of it I have to add heptagon and for every heptagon I must add an extra pentagon. Do you know any simple rules so that I can actually make a large sphere?

P. W. Fowler (*Exeter University, U.K.*). I believe you require 60 heptagons and 60 extra pentagons because they have a plane of symmetry, if you are going icosohedral, and you can do it with a slightly modified spiral algorithm.

T. Tarnai. We don't deal with dynamics; it was a static calculation. We considered elastic bars, which are usually in sub-stress. We found the equilibrium configuration, under which the structure has a certain shape, and in this case we can find only the coordinates of the proper nodes. We found the planes of the curved members which are circular or curved, and whether or not you have twisted or bent elements. In this case the member forms a helix too, but we can make no further comment at this stage.

P. W. Fowler. I was thinking of the actual moment from your initial guess as to the structure towards your bent and bulging final structure. So it is a process in time, but not actually dynamic. Furthermore, there is a chemical corollary of this description of the domes, and the rigidity of a structure with triangles, and that is that there is an extention to Euler's theorem using symmetry. This says that the symmetry spanned by the internal coordinates of a deltahedron is the same as the symmetry spanned by the edges. What this means is that any deltahedron skeleton cannot vibrate if it has perfectly rigid bonds. There are no true bending vibrations in deltahedrons, so there is a kind of chemical equivalent to your description.

T. Tarnai. So what can I say about a triangulated sphere is that for a long time mathematicians believed that there was no such structure which had free motion which would constitute a mechanism. One of them proved that such polyhedra do not exist, but Bill Conolly (late 1970s) produced a triangulated sphere which has free motion. This meant that it had a mechanism involving finite motion so if you wish to consider it as a structure that can vibrate, it has motion that causes no stress. Such a structure exists. Later a German mathematician produced another, so we have some examples.